The Future of the Artificial Mind

Alessio Plebe
Department of Cognitive Science
University of Messina, Messina, Italy

Pietro Perconti
Department of Cognitive Science
University of Messina, Messina, Italy

CRC Press
Taylor & Francis Group
Boca Raton   London   New York

CRC Press is an imprint of the
Taylor & Francis Group, an **informa** business

A SCIENCE PUBLISHERS BOOK

First edition published 2022
by CRC Press
6000 Broken Sound Parkway NW, Suite 300, Boca Raton, FL 33487-2742

and by CRC Press
4 Park Square, Milton Park, Abingdon, Oxon OX14 4RN

© 2022 Taylor & Francis Group, LLC
CRC Press is an imprint of Taylor & Francis Group, an Informa business

Library of Congress Cataloging-in-Publication Data (applied for)

ISBN: 978-0-367-63826-9 (hbk)
ISBN: 978-0-367-63827-6 (pbk)
ISBN: 978-1-003-12086-5 (ebk)

DOI: 10.1201/9781003120865

Typeset in Times New Roman
by Radiant Productions

To Oscar,
to Minou,
and to Chico.

Preface

Over the last decades, cognitive science has gained tremendous prestige. Thanks to it, the way we live our daily lives is changing once again. Jobs are changing as they are increasingly influenced by the ability to process digital information and are increasingly dependent on automation and robotics. Our psychological abilities are also becoming more powerful thanks to a number of new cognitive prostheses, including mobile phones and social media. Thanks to achievements in the field of autonomous driving, we can hope that traffic in our cities will one day be a completely safe experience. These are just a few of the areas where everyday life is changing at an accelerating pace due to the pressures of cognitive science.

After the first turbulent decades, in which the initial leadership of computer science was gradually disputed by linguistics, philosophy and psychology, neuroscience prevailed in the end. Thus, to this day, it is the brain sciences that have taken the lead role for cognitive science. From the Decade of the Brain (1990–2000) to the major projects simulating brain function, such as the Brain Activity Map Project and Brain Research through Advancing Innovative Neurotechnologies (BRAIN), the scientific community felt it could finally decipher the inside of the "black box" and thus simulate its functioning. The prospects for rehabilitation, for cognitive enhancement, and for artificial intelligence modelled on the neurocognitive architecture of the human brain sounded very promising.

However, things did not turn out quite as expected. The neuroscience discoveries were undoubtedly very significant, but did not quite live up to expectations. In the meantime, artificial intelligence has experienced something of a new renaissance. Ironically, this renaissance has not come about—as one might have expected—through the implementation of human-like cognitive architectures, but thanks to the mathematical sophistication that has enabled the surprising results of Deep Learning techniques. The impressive development of these techniques was initially driven by engineering purposes, but their application to cognitive processes proved to be extremely productive. After a period of stagnation that lasted a couple of decades, machines have again begun to surpass the cognitive abilities of humans. Artificial intelligence is now back to reclaim its role as the hegemonic discipline in the conflicting landscape of contemporary cognitive science.

Where will this lead? This book attempts to provide an answer to this question by summarising recent achievements from the various fields that make up cognitive science, with a particular focus on the challenges of the artificial mind. There are still many areas where there are more questions than answers, such as concerns about killer robots, sexbots, and the ethical algorithms in self-driving cars. Artificial intelligence promises to improve our capabilities and contribute to a more prosperous and fairer society. To achieve these goals, however, it is important to guess the trajectory the future of the artificial mind will take. This is precisely what the following path is about.:

Contents

Introduction

This book is devoted to the artificial mind. In particular, it is devoted to its recent developments and its near future. Among recent advances, the new techniques known as deep neural networks and the current trend of artificial intelligence in embedded structures, especially in the field of humanoid robotics, are particularly addressed. As for the near future of the artificial mind, this should be addressed with great caution. Much of what we know about the (artificial) mind is conjectural in nature, and of course guessing the future is an activity that is most likely to fail. Nonetheless, the trends that are now emerging seem to indicate a near future in which intelligence will be distributed everywhere, but it will be difficult to identify its bearer. Put another way, it seems that we are heading towards widespread yet accountability-free intelligence.

It is not easy to say whether such a scenario is a truly desirable outcome. In any case, it is very different from what we are used to. In other words, the scenario seems very un-ecological. In our ordinary experience, it is people who are more or less intelligent. The mind is a quality we ascribe to people based on certain characteristics in their behavior. But when the mind is detached from its regular bearer, it becomes something quite different. It becomes a property of things. Here, the possibility of panpsychism, or rather, artificial panpsychism opens up. It is definitely a dystopian scenario, to use a term employed in many of the film and television series that have become fashionable in recent years. But, in fact, we need to be very careful. While imagining the future is rarely a satisfying exercise, preparing for the most likely scenarios is a social task of both artificial intelligence and philosophy.

The book is organized as follows: The first chapter plays the role of a theoretical introduction to the following argument. It aims to make explicit a number of prejudices related to the vocabulary with which we talk about the artificial mind. The idea is to examine the equipment in the toolbox that will subsequently be employed. Thus, we consider what the word 'mind' means, as well as the words 'computation', 'representation', 'artificial' and 'intelligence'. The second chapter is devoted to the theory of computation and its historical roots. It will show what is meant by thinking is equivalent to computation. We will see how this thesis has a side concerning implementation (Turing machine), and a side concerning the psychological mechanisms of attributing the property of intelligence to behavior (Imitation game). The third chapter is devoted to the so-called classical artificial intelligence, i.e., in the first decades after the Second World War, and its joint adventure with the incipient cognitive science. We will then trace the evolution of artificial intelligence towards more ecological and embodied models, until the decline of artificial intelligence as the leading discipline of cognitive science in favor of neuroscience, which has since gained great prestige in the contemporary scientific landscape.

Before looking at the return of artificial intelligence as a leading discipline in cognitive science, chapter four surveys the major contributions that philosophy has made to the development of artificial intelligence. The classical opposition between rationalism and empiricism will, of course, play a leading role. But the chapter will also look at some of the other theoretical approaches that have influenced the development of artificial intelligence, from the theory of natural selection to inferential statistics, to the curious connections between computers and Heidegger's philosophical theories. Special emphasis will be placed on the role of language modeling, both in terms of how natural languages work and how artificial languages are modeled. The history of artificial intelligence is influenced by many variables but, overall, it is language that plays the role of the most important reference model.

The fifth chapter is devoted to a kind of renaissance of artificial intelligence. After a relatively stagnant period, artificial intelligence has regained the role of the leading discipline in cognitive science. This is largely thanks to the wonders of Deep Learning. While the latter has primarily evolved from engineering intentions, it has ultimately taken on the role of a game changer in the field of cognition as well. Machine translation, data security, and visual recognition are just a few of the areas where deep neural networks have forced scientists to change the direction and methods of their research. Understanding the role of deep neural networks in knowledge modeling is now, therefore, an unavoidable task for anyone who wants to understand the direction of the artificial intelligence adventure.

The sixth and seventh chapters are devoted to some moral, social, and philosophical problems made urgent by the recent development of artificial intelligence. Increasingly close relationships with machines equipped with artificial intelligence challenge common sense. Is it possible to fall in love with a robot? Is it ethical to have sexual relationships with such machines? If cars really will become self-driving vehicles, how should their behavior be modeled when faced with a decision that has ethical significance? And, most importantly, is a scenario in which machines can acquire some form of consciousness and self-consciousness credible, and what does it mean?

Attempting to answer these questions will lead the reader into an exercise of imagination regarding the future of the artificial mind, which is really the purpose of this book. Soul seems to be an outdated concept in such a scenario. However, the word 'soul' helps us realize that what we are talking about is not just one of the many technological achievements of modernity, but something capable of radically changing the way we think about human nature. That is why the book concludes with a section on the souls of the future and on the dystopian scenario in which intelligence in its artificial variant is enhanced and spread everywhere, including wearable technologies and the Internet. However, unlike its natural variant, artificial intelligence loses its bearer as it spreads everywhere. With the loss of the bearer of intelligence, we also risk losing who is responsible for the intentional states we are inclined to consider as intelligent. The dystopian outcomes of such futuristic scenarios should alert us against over-enthusiasm about the future of the artificial mind, and help to develop a balanced and constructive attitude towards AI and its impact on contemporary society.

Chapter 1

The Landscape

1.1 Into Eega Beeva's Toolbox

If you are a millennial, you probably take for granted things that were futuristic to those born before the 1980s. You probably assume that when you consult a news website, it comes in real-time from all over the world. You take it for granted that, with a twenty-hour or so flight, any corner of the planet can be reached. You take for granted that, with your cell phone, you can talk to, see, and exchange messages with people anywhere. We take things for granted that were just science fiction scenarios after World War II. If you are not a millennial, as a kid, you may have fantasized about strange drawings where you could see flying cars, cities floating in the air with thousands of ethereal streets running through them, fantastic glasses that let you see the underwear under girls' dresses, guns that could hit their target with laser beams capable of disintegrating any object in an instant. None of this fantastic world created for the imagination of boys after World War II was then realized as expected. But we knew that, after all. Nor, by the way, did we expect the cowboy world of the past to take the very form with which we entertained our imaginations at the time.

On the other hand, the future held for us just the sort of surprises we fantasized about as kids, though perhaps not quite in those areas. Global and digital communication, high-speed transportation, armaments, technological medicine are just a few of the achievements that have made many people's lives almost futuristic. We live twice as long as our ancestors and, if we go to a hospital, we are investigated by machines that can literally see inside our bodies. After all, the dirty nerd glasses have been replaced by something more useful. There aren't the laser guns we imagined, but the military launches attacks thousands of miles away via airplanes that fly without pilots and hit their targets from heights of thousands of feet without being seen by anyone. Cars don't yet fly, except in a

Figure 1.1: Eega Beeva by Bill Walsh and Floyd Gottfredson, from *The Man of Tomorrow*, 1947.

few prototype cases, but they now travel alone, gradually transforming our role from driver to passenger.

Few things have played a greater role in shaping the contemporary imagination than Walt Disney's characters and stories. This is also true in the realm of imagining the future. In a 1947 issue of Mickey Mouse, a character appears who is destined to affect the most common way of imagining the future and the role that science would play in shaping it: Eega Beeva (proper name: Pittisborum Psercy Pystachi Pseter Psersimmon Plummer-Push). Pluto and Mickey Mouse are forced by bad weather to take refuge in a large cave where they meet a strange individual who seems to come from the future. In fact, the title of Mickey Mouse's issue was just that: *The Man of Tomorrow* (Walsh and Gottfredson, 1947). Eega Beeva is not entangled in the quarrels of humans. He has a big head, a disarming confidence in the future and a bunch of extraordinary tools to pull out whenever needed. Taking a peek into Eega Beeva's toolbox would be very useful to embark on the journey that this book proposes to its reader.

In this book, we would, in fact, like to investigate what form the mind takes as we go deeper into the future. We would like to show how cognitive science is leading us to conceptualize the mind as something profoundly different from what we expected. But, more importantly, we would like to understand how the mind is transforming into something artificial. The future of the artificial mind is, therefore, the subject matter of this book. As we will see at the end of the path, it seems that the distribution of the mind in the social environment and the loss of a subject that owns it are the main features of the transformation to which intelligence is subjected in the historical phase in which we are living. Perhaps we will have to live in a world in which intelligence will be everywhere, but there will be no one to own it. It is too early to say whether a world characterized by distributed intelligence, but with no one in charge, will be better than the current one. After all, these are only predictions and, as happened in the case of X-ray glasses and fMRI, things in the future will take an amazing shape, but not exactly in the sense we had predicted, or hoped.

If we are interested in the shape that human and artificial minds will take in the future, casting a glance into Eega Beeva's toolbox would be of great help. Put another way, what is the right vocabulary to tackle the artificial intelligence journey? How should we use the words 'mind', 'artificial', 'intelligence', 'representations', or 'computation'? It is important to note the modesty with which this last question is inspired. It has, in fact, no foundational intent. This book makes no foundational claim. It is not like Whitehead and Russell's *Principia Mathematica*, in which the proof that $1 + 1 = 2$ does not appear until page 379. The minimal convention on vocabulary just mentioned has the sole purpose of making explicit that a book devoted to artificial intelligence and cognitive science necessarily includes a number of prejudices regarding the use of numerous expressions and words.

1.2 Mind

The word 'mind' refers to a set of cognitive capacities such as attention, perception, consciousness, imagination, computation, judgment, and language. How each of these capacities develops and establishes relationships with the others constitutes the shape of a particular mind. Hence, the mind of a rabbit will be different from that of a human being, that of a child from that of an adult, that of a robot different from the collective intelligence of the internet, whatever that latter expression may mean. Having a mind, however, does not only mean having some cognitive capabilities, but also experiencing the exercise of those same capabilities. Having a mind means experiencing perspective in the first person, having a subjective point of view. On the other hand, having a mind is the condition for having not only a personal identity, but also a social one. The social roles that we play in our lives are the result of complex social attributions based on the use of intentional vocabulary, made of 'beliefs', 'desires', 'emotions', and 'promises'.

The mental phenomenon is investigated from numerous scientific perspectives such as psychology, linguistics and artificial intelligence. Philosophy makes its contribution to the modern science of the mind primarily by attempting to answer the following three questions: What is it? Where is it? What is it for? The first question is a metaphysical one and aims primarily at establishing whether 'mind' is something that can be included among natural phenomena or whether, instead, it is an exception. The question about the place of the mind concerns the hypothesis that the mind is just a function of the brain, unless it is actually something more abstract. It is, therefore, a certain functional organization of matter, or if instead of trying to trace the place of the mind it is necessary to look elsewhere, i.e., in social relations, instead of in people's heads. The third and final question (What is it for?) alludes to the evolutionary advantage that having a mind must entail, given the enormous amount of energy that bodies devote to performing various cognitive tasks. Having a mind represents a significant evolutionary advantage especially when considered in relation to the development of language skills and social cognition.

1.3   Artificial

'Artificial' is a word that does not have a good reputation. It is the opposite of 'natural', which, by contrast, is a word that meets some of the prevailing sentiments in contemporary culture. 'Natural', in fact, refers to anything that is in harmony with the order of the world, from ecology to organic foods. Since contemporary man often feels guilty about the ecological footprint his actions leave on the world, anything artificial seems to allude precisely to man's inability to realize his own social development in harmony with the rest of the world. Yet 'artificial' also has a positive meaning, as it refers to the human ability to produce artifacts through its own intelligence. Technological artifacts, both intangible and cognitive, such as a website, and more concrete, such as a milling machine, are the pride of modern innovation and represent the main evidence of the importance of science in human development.

Technological artifacts are the natural outgrowth of the modern enterprise, driven by scientific knowledge and its vocation to influence social organization in democratic and participatory ways. But if intelligence itself is a candidate for becoming artificial, things change radically. The theoretical conundrum hinges on the fact that modern man sees scientific enterprise and the resulting social change as the result of the development of human intelligence. It is the humanistic project born in the sixteenth century that then animates the birth of experimental science and the Enlightenment that is the basis for typical modern optimism about the future. It is human intelligence that guides the whole process. In what sense could this intelligence become 'artificial' without betraying the original humanist project? The answer to this question rests on the confidence that any artificial intelligence should, nevertheless, be no more than an extension of the natural intelligence of man. In the humanistic project of modern science, artificial intelligence is something acceptable only insofar as it is conceived as a prosthesis of human intelligence. For this reason, all theoretical adventures perceived as liberating artificial intelligence from human intelligence are generally viewed with suspicion. Take, for example, Alan Turing's imitation game, the possibility of an evolution of intelligence that can take us to the Singularity stage, or John Searle's Chinese room. These are all cases in which we notice a kind of ideological resistance to accepting the very idea of 'artificiality' precisely because it seems to deviate from the path pointed out by natural intelligence and thus betray its own mission.

Artificial intelligence is therefore destined to be, at once, the most advanced fruit of the enterprise of modern science, but also the prime suspect for the perversion of its most original spirit. The advancement of artificial intelligence is a socially beneficial endeavor. This is evidenced by the AI4SG (Artificial intelligence for social good) research field, which is an attempt to use artificial intelligence to address social problems and improve the well-being of the world. But, at the same time, artificial intelligence is always seen as something that needs to be monitored and carefully controlled to avoid those apocalyptic scenarios typical of modernity, where machines end up controlling people, thinking is no longer free, and society is reshaped in an authoritarian sense and dominated by obscure technological forces.

1.4 Intelligence

Of all the words we can find in Eega Beeva's toolbox, 'intelligence' is the one characterized by the most general meaning. If we refer to the psychology of faculties developed over the centuries of modernity, which in turn is rooted in the way medieval and ancient philosophy had described the totality of human faculties of knowing the world, intelligence appears as a general term that includes all the various cognitive faculties of man. Perceiving, speaking, calculating, reasoning, persuading others: these are all ways in which intelligence is articulated in increasingly specific ways. Intelligence is what distinguishes human beings from stupid matter. It is what distinguishes res cogitans from res extensa and would attest to man's likeness to God.

The cultivation of intelligence is the main task of man because it distinguishes us from the rest of nature. Hence intelligence requires education and all the social structures like family and school that make it possible. This is why intelligence, if it is in danger of becoming something entirely artificial, is also in danger of betraying its deepest meaning. If intelligence is what makes us human, to make it artificial is to turn it against humans. This way of thinking is still deeply ingrained in many human cultures, especially in the West. But it has also been challenged by numerous scientific achievements. Particularly significant among these are those derived from studies of animal behavior. The Cartesian view of nature had clearly subsumed other animals into mere res extensa. Thus, it was imagined that there was the same qualitative difference between other animals and humans as there is between Dante's Divina Commedia and the stupidity of a stone we might trip over while walking. However, as the study of animal behavior has shown that animal behavior is guided by intelligence and reason just as much as human behavior, the qualitative difference underlying the Cartesian worldview has been strongly questioned.

The possibility that machines are endowed with intelligence is, therefore, one of the most typical adventures of modernity and experimental science. It is, in a sense, its vanguard. If you want to know where the trajectory of modernity is headed, you need to look closely at the trajectory of artificial intelligence. If you like the futuristic outcome of artificial intelligence, then you will like the foundations of modernity. If, on the other hand, the more distant and futuristic scenarios of artificial intelligence evoke a sense of fear and anxiety, then it is precisely the project of modernity that you dislike after all. In that case, you will be drawn to alternative theoretical scenarios, such as postmodernism, the New Age, or some sort of nostalgic return to a historical period in the more distant past.

1.5   Representations

One of the most popular metaphors in cognitive science goes back to Marvin Minsky (1986). It is the idea that the mind is like a society, a vast organized society, something like a city, in which everyone has a job to do and where, on the whole, it seems that everyone cooperates according to a common pattern. If the mind here resembles a city, then its inhabitants are the (mental) representations. The representations are the bearers of meaning because they always have content. And it is precisely because of this content that the mind is a device that does not wander aimlessly, but is something that has meaning. Representations are the bridge between the mind and the world. They are what enable the mind to talk about the world, to be able to relate to it. The mind is, of course, a part of the world. But in order for it to be able to refer to it, it must consist of representations (one could also say, be inhabited by them).

Representations are internal reproductions of particular aspects of the world. They involve the difference between something that is internal and something that is external. If you are a thinker in whose eyes the difference between internal and external, between subject and object, appears as something naive, then representations are not for you. Representations are typical creatures of modernity, a way of thinking that is trapped by the image of a subject trying to make a picture of what is in front of it. Without representations, there is no mind. Or at least there is no mind in the modern sense of the word. Otherwise, the mind can be understood as the 'breath of the world', so to speak, i.e., as an emergent property of the world. In this account, the mind is not regarded as a product of the subject, but as a property of the world itself. In other words, the mind is the way the world works. This is a sophisticated idea, one that might meet with the approval of great thinkers like Georg Wilhelm Friedrich Hegel. But it is not the idea on which cognitive science is based. The latter, on the contrary, is based on an idea typical of modernity, emphasized by philosophers like John Locke and Immanuel Kant, according to which the mind is an internal construction of a subject that tries to orient itself in the world by representing it in itself according to a certain logical form.

Representations, however, are not concrete objects. They are like numbers and the basic units of arithmetic. More precisely, mental representations are abstract rules. Their task is to establish a relation between two domains. They are comparable to the notion of mathematical function. A function in this context is a relation between two sets that can establish a correspondence between the elements of the first set, called domain, and the elements of the second set, called codomain. In the brain of an animal, a mental representation is like an abstract rule, even if it is physically realized in configurations that are present in the nervous system. The rule provides a link between behavioral types (domain) and environmental influences (codomain). Even if we are able to trace something in the brain that acts as a counterpart to a particular property of the world or object of the world, it is not a mental representation. It is a physical configuration that is connected to a particular class of environmental input via a mental representation. Thus understood, representation has the advantage of being a concept that can be used in a computational context while having a naturalistic counterpart in neural activations and in the class of environmental stimuli that can trigger them. For this reason, representation is the best theoretical construct available to a science of the mind that aspires to be both computational and naturalistic.

This idea of what makes a mental representation is particularly elegant and lightweight. It requires us to represent the phenomenon of knowledge through the naive notion of a subject attempting to create an internal model of the external world through sense systems and the brain. But in other respects it is a particularly liberal idea, that is, it is compatible with most of the theoretical attitudes that have characterized the new science of mind that arose after World War II. It is compatible with an abstract and purely computational idea, typical of the cognitive science of the first decades. It is compatible with a cognitive science that tries to be ecologically correct. And it is compatible with 4E cognition, that is, embodied, embedded, enacted, and extended cognition. Although some radical enactivist philosophers, such as Dan Hutto (1999, 2013) and Shaun Gallagher (2008, 2017), have stated that cognitive science should dispense with the very idea of representation, this is actually an overstated conclusion. They object to the idea that the mind consists of manipulations of explicit representations characterized by the propositional format. They also try to counter the view that knowledge has to do with the mirror metaphor, i.e., that it is essentially about creating a reproduction of the world. On the contrary, they think (rightly) that knowledge is primarily to do with the notion of 'action'. The mind is not a speculation of the world to which action is then eventually added, directed outward. On the contrary, thought is directed to action from the outset because the relation between the individual and the perceptual scene in which he is involved is an essential one. These are all valid considerations when turned against the notion that representations are internal objects aiming to statically map the world by means of explicit representational constructs. However, these are remarks indifferent to the minimal and elegant idea of representation we tried to illustrate earlier.

The term 'representation', moreover, is a generic term that encompasses many different kinds. There are explicit and propositional representations, which irritate the radical enactivists, but there are also implicit ones. Moreover, mental representations can take on the musical or visual format in addition to the propositional format. Mental representations are what guide us when we hum a particular tune in our heads, and it is mental representations as much as visual ones that we generate in our heads when we want to answer questions like the following: How many rows of strings are there on the spire of the Chrysler Building in New York? In addition, representations can be perceptual and articulated according to different sensory systems. They can also be motor, such as those that allow us to articulate a complex movement, as performed when walking or serving in tennis. If mental representations are the inhabitants of the city of the mind, then citizens can perform a range of very different functions. One can be a postman or a bus driver; likewise, one can be a motor representation or a perceptual one; one can be endowed with a symbolic format or a visual and figural one. Different skills and roles are needed for different tasks in the city.

In this way, we have a model of the mind as if it were a city inhabited by different kinds of citizens. But they are still stationary, each in their own uniform, according to the role each has to play once the social game of the city gets going. For this to happen, it is necessary that there be computations as well as representations. Computation is, indeed, the engine of the mind.

1.6   Computation

Associating computation with the mind is a risky theoretical move. As we shall see below, it is an idea that has a history going back centuries. Still, there is something unnatural about the idea that thought is a form of computation. The point is that computation is cold and impersonal, while the mind is expected to be something that expresses a first-person perspective as informed by emotion. This, basically, is the difference between first person perspective and third person perspective. The latter explains things as nature or God might see them. Having such a perspective is wonderful. It frees us from personal biases, from social biases that depend on the class to which we belong, as well as from the conceptual patterns of the culture to which we belong. It elevates us to the level of the things themselves. The chemical description of chlorophyll photosynthesis has the power to raise us above nature; it gives us the ability to describe it as God himself could. We can imagine that if God were to write a book on chemistry, he would do so just as we do.

If, at a certain point in the history of science, this possibility even affects our own inner lives, we are almost seized with a Promethean frenzy. We are not only no longer able to report objectively on how things are in the world, but even how we make up our own minds about it. Nothing has to be brought back into the frame of reference of experimental science anymore. Everything is now within the same explanatory mechanism. The distinction between Geistes- und Natur-wissenschaften, which had been in vogue in Germany since the nineteenth century to mark an unbridgeable gap between the two substances of the world, has at last been swept away. We behold at last the theory of everything, a single explanatory instrument capable of providing an explanation for every possible phenomenon, from the photosynthesis of chlorophyll to the reason why the scent of certain cookies is able to elicit just that particular course of emotion in our individual minds. This is such an ambitious perspective that it certainly seems to be the result of a certain human arrogance. It is as if Icarus had found the right wings to fly into the sky without being scorched by the enormous heat of the sun.

To make the idea so Promethean that thought is nothing more than a form of calculation, another consideration that is just as attractive as the previous one is added. Just as it seemed impossible to bridge the gap between the first-person perspective and the third-person perspective, it seems as impossible to bridge the gap between intelligence and stupid matter. However deeply rooted the naturalistic stance pushing to bring all phenomena for which we seek an explanation into the framework of the natural sciences, there seems to be a qualitative difference between the human mind and a stone. Namely, a stone will never be able to ask itself the kind of questions with which this book is filled. Very well, stones are wonderful objects. The point is not to impose an onerous ontological hierarchy on the world based on the cognitive distortion that such an imposition comes precisely from us humans and not from stones. But, even if one adopts such a liberal ontological stance, one cannot get over the stark fact that being a stone proves nothing, and that there is a huge qualitative difference between inert matter and a living organism that can call itself into question.

Yet computational psychology promises to do just that. In a sense, philosophy of the mind is an anachronistic discipline. If it really consists in questioning the relationship between mind and body, then its very raison d'etre fails. Indeed, the miracle of the Turing machine and of computational psychology, with which we will deal in detail in the next chapter, seems to have rendered obsolete the categorical spasm on which philosophy of the mind tends to practice. Computational psychology is based on the idea that what appears intelligent to our eyes, what we are accustomed to call 'intelligence', is nothing but the result of a certain arrangement of inert matter. There is no unbridgeable gulf between inert matter and intelligence. The latter is just matter, arranged in a suitable way. It is obvious that there are many cultural prohibitions against seeing things in this way. But once those prohibitions are overcome, a new world opens up, consisting of previously unthinkable explanatory and technological possibilities.

Thus, the goal of cognitive science, which it shares with artificial intelligence, is to identify the correct cognitive architectures which are ultimately nothing more than computational architectures. The brain is no different than any other organ in this regard. The kidneys perform an essential role for the functioning of our body. For a long time, this role remained undiscovered and we were unable to explain how the kidneys functioned. However, when we discovered the secret, we found that it was a specific functional architecture. What we discovered was a particular mechanism, a set of configurations that are suitably connected. There is nothing that connects the matter that the kidneys are made of to their biological function. In fact, once we discovered the functional architecture of the kidneys, we were able to get artificial machines to perform that function in dialysis. It is the same with the brain and the mind. Having discovered that there is nothing to bind the soft matter of the brain to that of which human dreams are made, the adventure of artificial intelligence seems, in principle, to have no obstacles here. Just as the dialysis machine can work for a kidney, artificial intelligence can work for every aspect of the inner workings of the human mind. Of course, some problems will initially seem more difficult than others, such as consciousness and especially its qualitative side. Implementing visual recognition on a machine seems somewhat more plausible than building a machine that can experience the subjective sensation of consciousness and even the indescribable feeling of the effect of simply being who one is. But it seems only a matter of time and application. In the meantime, the veil has fallen. There no longer seems to be an unbridgeable gulf between first and third person explanations, and between stupid matter and intelligence.

The Promethean adventure of artificial intelligence can now begin. In the next chapters, we will try to describe this adventure and guess where the road may eventually lead us.

Chapter 2

When the Computer First Meets the Mind

The basic tools for working the artificial mind of the future, just introduced in the previous chapter, were forged in the past. In this chapter we trace the history of the idea of thinking as computation, back to the foresight of those such as Thomas Hobbes. The stage for the forthcoming computer was set by two distinct scientific efforts: the development of logic, and the invention of physical machines. In the first effort we found thinkers like Gottfried Leibniz, George Boole and Gottlob Frege. Leibniz has to be counted among the inventors of machines as well, together with others like Blaise Pascal and Charles Babbage. At the dawn of the computer age, the stage was dominated by Alan Turing. If there is a person in the world that can be referred to as the father of the computer, it is him. The two most celebrated contributions of Turing are his 1936 abstract machine, and his 1950 test of intelligence. The former is the foundation of computer science, and since AI is part of computer science, the Turing machine is of some significance to AI too, but only as a background. Conversely, the Turing test is considered to represent the very beginning of AI. Curiously, for cognitive science, the roles of the two main contributions by Turing have almost been reversed, with an earlier and larger influence of the Turing machine with respect to the Turing test. One of the first thinkers conferring mental citizenship to the Turing machine was Hilary Putnam. When AI comes of age, the Turing test, too, will become a central topic in cognitive science, as is the famous mental experiment formulated by John Searle in 1980—known as 'Chinese Room'—as counter-argument to the Turing test. Along this brief historical note we find, in 1956, the official birth of AI.

2.1 Thinking as Calculating

Metaphors have been ubiquitous and vital in every scientific discourse (McCormac, 1976; Brown, 2003). However, nowhere else in science, perhaps, has there been a greater reliance on metaphors than in reflections about the mind. This is hardly surprising; it is difficult to find something in the universe that is at the same time so powerful and complex, and so elusive, as the mind. Therefore any attempt at its explication had to resort to analogy with better known domains, especially before modern neuroscience. John Daugman (1990) has compiled a comprehensive historical catalogue of metaphors used for mind and brain theories.

Thinking as computing can be the explanatory metaphor for the mind that incorporates the most pervasive device of today—the computer - not very differently from the water technology of antiquity or the clockwork mechanisms during the Enlightenment (Vartanian, 1973). There are, however, key differences between computing and the previous metaphors of the mind. As introduced in §1.6, computing transcends mere metaphor, offering the Promethean ambition of being able to explain mental experience from a naturalistic third-person perspective.

As we will describe now, the idea of thinking as computing appeared in history well before the device today called 'computer', and refers to the abstract process of mathematical computation. Moreover, most proponents of this idea have interpreted it as almost a literal description of mental functions.

An early and peculiar form of computation as the correct basis of reasoning was already conceived in the Middle Ages by Ramon Llull (1310), not only an amazing mathematician, philosopher, but also an active crusader. His major work, *Ars Magna*, is a sort of combinatorial system that should be able to disclose the truth of any sort of rational debate. He developed this system mainly for settling disputes between infidels and Christians. Even if his credentials as crusader lead one to suspect that his method was biased, his aim was to substitute bloodshed and fighting with geometrical diagrams and combinatorial operations. Llull tried to promote his method in the intellectual circles in Tunisia, but with little success; he was incarcerated for a while and then, eventually expelled.

A more long-lasting contributions to the idea of thinking as computing was offered by the well known English philosopher, Thomas Hobbes. The strict parallel between thinking and mathematical operations is clearly stated in his magnum opus *Leviathan*:

> When man reasoneth, he does nothing else but conceive a sum total, from addition of parcels; or conceive a remainder, from subtraction of one sum from another: which, if it be done by words, is conceiving of the consequence of the names of all the parts, to the name of the whole; or from the names of the whole and one part, to the name of the other part. And though in some things, as in numbers, besides adding and subtracting, men name other operations, as mul-

tiplying and dividing; yet they are the same: for multiplication is but adding together of things equal; and division, but subtracting of one thing, as often as we can. These operations are not incident to numbers only, but to all manner of things that can be added together, and taken one out of another. For as arithmeticians teach to add and subtract in numbers, so the geometricians teach the same in lines, figures (solid and superficial), angles, proportions, times, degrees of swiftness, force, power, and the like; the logicians teach the same in consequences of words, adding together two names to make an affirmation, and two affirmations to make a syllogism, and many syllogisms to make a demonstration; and from the sum, or conclusion of a syllogism, they subtract one proposition to find the other. Writers of politics add together pactions to find men's duties; and lawyers, laws and facts to find what is right and wrong in the actions of private men. In sum, in what matter soever there is place for addition and subtraction, there also is place for reason; and where these have no place, there reason has nothing at all to do.

Out of all which we may define (that is to say determine) what that is which is meant by this word reason when we reckon it amongst the faculties of the mind. For reason, in this sense, is nothing but reckoning (that is, adding and subtracting) of the consequences of general names agreed upon for the marking and signifying of our thoughts; I say marking them, when we reckon by ourselves; and signifying, when we demonstrate or approve our reckonings to other men.

(Hobbes, 1651, p.c.V)

The entities, called 'parcels' by Hobbes, on which operations like addition are performed, play the role of tokens composing thoughts, corresponding roughly to what in later cognitive science are usually called 'symbols'. The equivalence between thinking and computation is neatly stated in the later *De Corpora*:

By reasoning I understand computation. And to compute is to collect the sum of many things added together at the same time, or to know the remainder when one thing has been taken from another. To reason therefore is the same as to add or to subtract.

(Hobbes, 1655, p.1.2)

Such a commitment of Hobbes for a computational interpretation of the mind has led Haugeland (1985, p.23) to call him 'the grandfather of AI'.

2.1.1 Logic

Hobbes never attempted to elaborate the mathematics behind reasoning; it was neither among his intentions nor his possibilities, considering he did not master mathematics. The proper way to describe thought in mathematical terms would be found later, in logic. Since Aristotle (335–323 BCE), logic was the attempt to make explicit the rules underlying human thought and language, but the rules were not expressed in mathematical form.

The progressive shift of logic towards mathematics has been fundamental not just for the computational account of the mind, but for the development of the digital computer itself. Martin Davis (2000) has recounted the history of logical concepts underlying modern computer science, culminating in the Turing machine, here described in §2.2.1. Davis starts his journey long before, with the great German mathematician and philosopher Gottfried Leibniz. He dreamed that a specific kind of calculus would be the key to settling all human conflicts and disagreements. In his words:

> Quo facto, quando orientur controversiae, non magis disputatione opus erit inter duos philosophos, quam inter duos computistas. Sufficiet enim calamos in manus sumere sedereque ad abacos, et sibi mutuo dicere: calculemus!

(Leibniz, 1684, p.200)

For this exhortation to be feasible a new mathematics was necessary, one he called *calculus ratiocinator*. His desired mathematics of thinking was conceived as an external tool to accomplish exactly any type of reasoning, and did not explicitly entail that our mental way of reasoning was mathematical in its essence, as it was for Hobbes. But unlike Hobbes, Leibniz was highly gifted in mathematics, and he actually planned to develop the *calculus ratiocinator* as the last effort of his life. Unfortunately he wasn't able to spend much time on this project, and he produced only a few preliminary fragments.

Two centuries later, the first mathematics of reasoning was laid down by George Boole (1854). His greatest work was in using standard algebra, giving variables and operations a new meaning related to mental thinking. Variables, for which Boole used the higher alphabetic letters, like $x, y, z, \ldots$, denote sets of objects satisfying a specific concept; for example x might be the set of `animal` and y the set of `green` entities. The product operation, whose correspondent symbol is omitted as in ordinary algebra, corresponds to the intersection set operation $\cap$, so that xy, in our example, is the set of green animals like frogs and lizards. The $+$ operator corresponds to the union $\cup$. There are two possible constant values: 1 corresponding to the universe of objects, and 0 to the empty set, therefore $1 - x$ is the set of all non animated objects.

The basic operations are ruled by the following set of basic laws:

$$xy = yx \tag{2.1}$$

$$x + y = y + x \tag{2.2}$$

$$z(x + y) = zx + zy \tag{2.3}$$

$$x = y \Rightarrow zx = zy \tag{2.4}$$

$$x = y \Rightarrow z + x = z + y \tag{2.5}$$

$$x^2 = x \tag{2.6}$$

$$x + x = x \tag{2.7}$$

Table 2.1: The three primary propositions in Boole's algebra. The second column from the left is the general format of the proposition, the third column is an example of the proposition, with its algebraic translation in the rightmost column.

universal	$f_S = f_P$	computer scientists are animals with keyboards drinking coffee	$p = ab(1-s)$
particular predicate	$f_S = v f_P$	computer scientists wearing glasses are nearsighted	$op = vm$
particular subject and predicate	$v f_S = v f_P$	some computer scientists with age become philosophers	$vtp = vf$

symbols used in the examples:	a = animals b = with keyboard f = philosopher m = nearsighted o = with glasses p = computer scientists s = coffee drinker t = with age v = the indefinite class

where $=$ is the identity symbol. Equations (2.1) and (2.2) are commutative properties, equation (2.3) is the associative, and (2.4)) (2.5) are identity properties. Equation (2.6) is called the dual property, and it is at the core of the deductive system proposed by Boole. For example, from (2.6) derives that $x(1 - x) = 0$, as in our example, that nothing can be an animal and not an animal at the same time. Equation (2.7) is the second dual property, and is exactly the same as the second Axiom of the initial fragments of Leibniz's *calculus ratiocinator*, seemingly not known to Boole. While the first dual property matches the equivalent algebraic expression, since $x^2 = x$ when x has only 0 and 1 as possible values, the second dual property is obviously different, for ordinary addition $1 + 1 \neq 1$.

Boole went further in relating algebraic expressions with propositions of natural language; for this purpose he introduced a special variable, v, the indefinite class. This was, in fact, Boole's expedient to express *quantification*. His three 'primary' propositions are those listed in Table 2.1, where f_S is an arbitrary expression denoting a subject, and f_P is an arbitrary algebraic predicative expression.

Note the use of v as surrogate for quantification in particular propositions. The logic system of Boole is completed with an elaborated methodology for actually 'solving' systems of equations, corresponding to propositions. It is divided into three main phases: elimination, reduction, and development. The first two are direct extensions of the ordinary methods of algebraic manipulations, like identification of superfluous variables and their elimination, the reduction of a system of equations to the minimum number. What Boole calls 'development' is, instead, specific to the meaning of the symbols in his logical system.

Analyzing this system in depth is out of the current scope of this book, but what is remarkable is that one of the main objectives Boole had in inventing his system was to describe the mental processes of reasoning. It is this aspect of his work that was completely removed in the ensuing developments of logic. Bertrand Russell (1918), in declaring his admiration for the pioneering work of Boole, alluded to his extravagance in connecting logic and mind: "Pure Mathematics was discovered by Boole in a work which he called *The Laws of Thought*."

Boole opened the road to logic, but his system was constrained in making use of symbols and tools inherited from ordinary algebraic calculus, which was invented for working with numbers and, despite his great efforts, was unmalleable for new purposes. Contemporary logic is mainly due to Gottlob Frege, and one of his first ideas was to invent, from scratch, a way of formally expressing 'concept'. For this reason, he named his system *Begriffsschrift*[1] (Frege, 1879). We can anticipate, right away, that nobody has ever used the *Begriffsschrift* after him, but nonetheless, some of the constituents of his system became the foundation of contemporary logic.

The *Begriffsschrift*, in its neglected aspect, is a curious new way of writing that breaks the left to right and top to bottom sequence common to western languages (and mathematical writing too). It develops in two dimensions. The basic form is the assertion: ├────── A, A is true. Unlike in Boole's system, variables like A are now propositions, sentences which can be considered as true or false. The two basic connectives are the negation and the implication, drawn as in the following, with the current notation below:

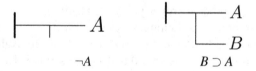

in every logical system, it is possible to define axiomatically two connectives only, and all the remaining can be derived; here, for example, are the conjunction and the disjunction in the pictorial *Begriffsschrift* form:

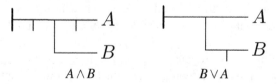

---

[1]Literally, a language for concepts.

The breakthrough of Frege's system is in two radical innovations with respect to Boole: the function and the quantification. The idea of function is borrowed from calculus, though with subtle differences (Dummett, 1973). Function and its argument, in Frege's system, correspond to any possible decomposition of a simple proposition into two components. Unlike in calculus, functions return only truth values. Here is how the combination of function and universal quantification appears in the *Begriffsschrift*; on the right is the current notation:

$$\vdash\!\!\!\underset{\smile}{\quad a \quad} \Phi(a) \qquad\qquad \forall a\,(\Phi(a))$$

The domain of quantification is expressed by the small basin where a is housed.

For example, the following complex statement:

`if all graybeards are talented, and Marvin is a graybeard, thus Marvin is talented` which, in contemporary logic, would be written in this form:

$$(\forall x\,(g(x) \supset t(x)) \wedge g(M)) \supset t(M) \tag{2.8}$$

would, in the *Begriffsschrift*, become:

$$
\begin{array}{ll}
t(b) & \\
t(a) \quad g(b) & \supset \\
g(a) \quad (\forall a\,(g(a) \supset t(a)) \supset t(b)) & \\
g(b) &
\end{array}
\tag{2.9}
$$

in which Marvin is *b*, to adopt Frege's convention on variable names; on the right there is the contemporary transcription of the (2.8), without using the conjunction.

Frege's project was ambitious, he wanted to base arithmetic upon logic, and, in addition to the *Begriffsschrift*, he developed a full axiomatic system (Frege, 1884). This beautiful system is corrupted by the excess of freedom in defining functions. The possibility for an argument to be itself a function has dramatic consequences, as discovered by Bertrand Russell, who posed the famous question:

> You state that a function, too, can act as the indeterminate element. This I formerly believed, but now this view seems doubtful to me because of the following contradiction. Let *w* be the predicate: to be the predicate that cannot predicate of itself. Can *w* be predicated of itself? From each answer its opposite follows. Therefore we must conclude that *w* is not a predicate. Likewise there is no class (as a totality) of those classes which, each taken as a totality, do not belong to themselves. From this I conclude that under certain circumstances a definable collection does not form a totality.
>
> (Russell, 1902, pp.124–125)

The failure of his project broke Frege's heart, he added an appendix to the *Grundgesetze* with an *ad hoc* limitation to the freedom in defining functions forbidding the argument of the function to be a function that has itself as argument. It was clearly a desperate way out (Quine, 1955), and in fact Frege never recovered from this blow. Nevertheless, the work of Frege paved the road of contemporary logic. Russell was one of the first to develop a logic system meeting the expectations of Frege to reduce arithmetic to pure logic without this paradox. His effort took the form of the *Principia Mathematica* (Whitehead and Russell, 1913), an opus regarded with veneration by future AI, as we will see in §2.3.2.

2.1.2 A Steam Powered Computer

This book is not concerned primarily with the physical aspect of computers, which is usually called the *hardware*. The tangible availability of computers is clearly the necessary prerequisite for the existence of AI and also of cognitive sciences; but the computational account of the mind is related much more with the intangible side of machines, which is usually called the *software*. Computer science itself was founded on a pure abstract idea, the Turing machine, that will be introduced in §2.2.1. Therefore, while our historical account of the development of concrete computers is condensed in this short section, the interested reader is referred to other books (Goldstine, 1972; Randell, 1973; Ceruzzi, 2003; Bauer, 2010; Dasgupta, 2014).

Ever since the dawn of mathematics, inventors have tried to make instruments to help in doing calculations. We skip all the ancient history, and move on to an outstanding figure we already met: Leibniz. With his extraordinary eclecticism, in addition to philosophy and mathematics Leibniz also engaged in refined design and mechanical construction, producing in 1671–1674, a machine that could perform the four basic operations of arithmetic. It was a significant evolution over the *Pascaline*, a machine built by Blaise Pascal in 1645 that could do addition and subtraction. Leibniz was motivated in his effort by scientific aims, to ease the work of the astronomers, while Pascal had more mundane goals, to help his father in the task of reorganizing the taxes of Brittany.

The transition from calculating machines to computers was first planned by Charles Babbage (1889). He was not only a fine English astronomer and mathematician, but was also highly gifted in mechanical design. He was probably not as brilliant in public relations; despite decades of work, none of the machines he conceived was built in his lifetime, largely because of funding problems and clashes of personality between Babbage and his stuff (Collier and MacLachlan, 1998; Swade, 2001). He planned a first steam powered machine, called *Difference Engine*, in 1820, able to compute polynomial functions up to the sixth order. The name 'difference' derives from the divided differences method for interpolating polynomial coefficients. By the time the government abandoned the funding of the project in 1842, Babbage was intrigued by a new and more advanced

Figure 2.1: The unfinished Analytical Engine of Charles Babbage (Science Museum, London).

machine. This machine, called *Analytical Engine*, would execute sequences of arithmetic and logical operations, with the possibility of changing the sequences conditionally based on partial results. These features would allow the automatic solution of a large class of analytical equations and the computation of complex mathematical expressions. Figure 2.1 shows the mill and the printing mechanism of the unfinished Analytical Engine. From 1995 to 1991 the Science Museum of London, under the direction of Doron Swade (2001), completed the construction of the Difference Engine, using the original design drawings of Babbage. The machine, made of 8,000 parts for a total weight of 5 tons, was the only steam powered working computer. While the construction of Babbage's machines during his lifetime was a failure, his efforts had an important impact on the burgeoning field of computer science, even on the software side.

As mentioned above, Babbage had difficult relationships with his collaborators, with one important exception: Lady Ada Lovelace, daughter of Lord Byron. She compiled a written description of the Analytic Engine (Lovelace, 1843), including her original ideas about the *programming* of the machine, exceeding Babbage's own vision for his machines, and anticipating key concepts of modern computing (Essinger, 2014). In order to turn the Analytic Engine into a universal machine, able to compute arbitrary analytic expression, the specifications of its procedure should take the form of an *algorithm* indicating every step of the procedure in some symbolic format. Ada Lovelace wrote one code for the Analytical Engine to compute Bernoulli numbers in this format, the first ever piece of computer software. When, in the 1970s, the US Department of Defense planned the development of a new programming language with the highest level of reliability and safety, adequate for critical applications such as military, banking, and space technology, the name Ada was given to the programming language (Ichbiah et al., 1979). The standard reference manual of the language was approved on the date of Ada Lovelace's birthday (December 10 1980), with the code number MIL-STD-1815 in honor of her birth year. In her Notes, Ada Lovelace (1843) pushed her vision even further:

> It [the Analytical Engine] holds a position wholly its own; and the considerations it suggests are most interesting in their nature. In enabling mechanism to combine together general symbols in successions of unlimited variety and extent, a uniting link is established between the operations of matter and the abstract mental processes of the most abstract branch of mathematical science. [...]
>
> We are not aware of its being on record that anything partaking in the nature of what is so well designated the Analytical Engine has been hitherto proposed, or even thought of, as a practical possibility, any more than the idea of a thinking or of a reasoning machine.

This passage appears to raise the possibility that a machine like the Analytical Engine may indeed be a sort of 'thinking or reasoning machine', even if Ada Lovelace stops somewhat short of such a heretical claim (Green, 2005). Indeed, she stepped back from this dangerous conjecture, with a subsequent note:

> It is desirable to guard against the possibility of exaggerated ideas that might arise as to the powers of the Analytical Engine. [...] The Analytical Engine has no pretensions whatever to originate anything. It can do whatever we know how to order it to perform.

As we will see soon (§2.3.1), a century later, Turing would dub this passage the *Lovelace Objection* to a thinking machine.

2.2   The Turing Machine

Turing followed somehow the stereotype of a mathematical genius, with his complex and troubled personal life. However, he was certainly not vain and conceited to the point of calling his main ideas *Turing machine* and *Turing test*. Those names come after him, and his famous test, in the original paper, spells *Imitation game*. This is the title of the recent (2014) award-winning film by Morten Tyldum that made Turing popular to a wide audience. Despite its title, the film has little bearing on the Turing test; its plot is all about Turing's great struggle in cracking the Nazi Enigma code in World War II.

Thanks to the popular success of this film, we can dispense with relating how much we owe to Turing for the fact that Europe could escape Nazi domination; we refer the interested reader to Turing's beautiful biography by Andrew Hodges (1983), and other historical essays (Aldrich, 2010; Ferris, 2020). In this section, we address the two celebrated papers of Turing that paved the road to computer science and AI.

2.2.1 The Machine

The first paper, written when Turing was a 24 year old student, is titled *On Computable Numbers, with an Application to the Entscheidungsproblem* (1936). Nowadays, English is the language of science, mathematics included. At the beginning of last century, Germany was a kind of center of gravity in mathematics (Siegmund-Schultze, 1997), charged by the weight of great mathematicians like Euler, Gauss, Weierstrass, Riemann. Therefore, it was not unusual to leave German terms untranslated, as Turing did for *Entscheidungsproblem*. Literally 'decision problem', it is an essential element of the program proposed by David Hilbert to ground all parts of mathematics to a unified set of consistent axioms. Let Hilbert himself, together with Wilhelm Ackermann, explain what this problem is:

> Das Entscheidungsproblem ist gelöst, wenn man ein Verfahren kennt, das bei einem vorgelegten logischen Ausdruck durch endlich viele Operationen die Entscheidun über die Allgemeingültigkeit bzw. Erfüllbarkeit erlaubt. [...] Das Entscheidungsproblem muss als das Hauptproblem der mathematischen Logik bezeichnet werden[2].

(Hilbert and Ackermann, 1928, p.73)

Turing became intrigued by the *Entscheidungsproblem* in 1935, while attending a course on the foundation of mathematics held by M.H.A. Newman at King's College, Cambridge. He was convinced that the response to Hilbert must be negative, the deciding procedure cannot exist; but in order to demonstrate that, he would have to invent a strict mechanical process able to analyze any given logical expression. Turing created an ingenious procedure that, despite being dismaying simple, is so general as to process any mathematical problem. This procedure was soon dubbed *Turing machine* by Alonzo Church (1937), in his review of Turing's article. No one could have felt a greater interest in Turing's article than Church because, in the same year, he produced his own strategy, the $\lambda - calculus$ (Church, 1936), to solve the *Entscheidungs problem*. In the same period Emil Post (1936), independently found yet another general mathematical mechanic procedure. From a theoretical point of view, the formulations of Turing, Church, and Post are perfectly equivalent, but while Post and Church remained in the exclusive and rarefied world of abstract mathematics, Turing, with his machine, was bound to become the father of computer science (Daylight, 2015).

Informally, the Turing machine, sketched in Figure 2.2, consists of a memory tape divided into cells, on each of which a symbol can be read and/or written, and a state–transition machine, programmed by a set of transition rules. The machine

---

[2]The *Entscheidungsproblem* is solved when we know a procedure that allows, for any given logical expression, to decide by finitely many operations its validity or satisfiability. [...] The *Entscheidungsproblem* must be considered the main problem of mathematical logic.

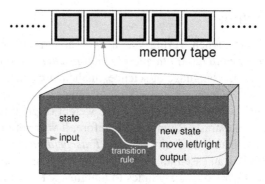

Figure 2.2: Sketch of the Turing machine.

Table 2.2: Correspondence between the terms used by Turing in his original article and those currently used for components of the Turing machine.

current term	Turing's original term
Turing machine	*a*-machine (*a* for *automatic*)
state	*m*-configuration
cell	square
configuration	complete configuration
transition	move
infinite-loop	circle-free
not infinite-loop	circular
universal Turing machine	universal computing machine

scans just one cell of the tape at a time, and proceeds to scan the cell immediately to the left or to the right of the current cell, depending on the transition rule.

For a formal definition, we will use the mathematical language commonly used in contemporary computer science; for the corresponding original terms used by Turing, see Table 2.2. A Turing machine, M, is the following quintuple:

$$M \overset{\text{def}}{=} \langle Q, A, \delta, q_0, F \rangle \qquad (2.10)$$

where Q is the set of *states*, A is the alphabet of *symbols*, $q_0 \in Q$ is the initial state, $F \subseteq Q$ are the final states, and δ is the transition function. It is a partial function, defined as follows:

$$\delta: \ (Q - F) \times A \to Q \times A \times \{\Leftarrow, \Rightarrow, -\} \qquad (2.11)$$

With today's insight, it is impossible not to see the transition function δ as a computer program, or software. Given a state $q \in Q$ and a symbol $a \in A$, the application $\delta(\langle q_i, a_i \rangle)$ is actually a rule that instructs the machine on what to do. And what the machine can do is extremely simple: change its status, possibly

Table 2.3: Example of a Turing machine that keeps cat and dog in order. The table with all instructions is on the left, note that in the output column the mark "–" means that the current symbol is not modified. On the right the machine is executed step by step on an example tape, starting from the top. The leftmost number is the state of the machine at every step. The final tape is at the bottom, when the machine has reached the final state 4.

STATE	INPUT	NEXT STATE	CELL	OUTPUT
0	dog	0	⇒	–
0	cat	1	⇒	–
0	#	4	-	–
1	dog	2	⇐	cat
1	cat	1	⇒	–
1	#	4	-	–
2	dog	3	⇒	–
2	cat	2	⇐	–
2	#	3	⇒	–
3	–	0	⇒	dog
4	–	-	-	–

Execution on example tape:

	⇓						state
#	dog	dog	cat	dog	#		0
#	dog	dog	cat	dog	#		0
#	dog	dog	cat	dog	#		0
#	dog	dog	cat	dog	#		1
#	dog	dog	cat	cat	#		2
#	dog	dog	cat	cat	#		2
#	dog	dog	cat	cat	#		3
#	dog	dog	dog	cat	#		0
#	dog	dog	dog	cat	#		1
#	dog	dog	dog	cat	#		4

move one cell to the left or to the right, and possibly change the symbol in the current cell.

At every step i of the computation, the machine is fully described by its *configuration*, the following triple:

$$C_i \overset{\text{def}}{=} \langle q_i, L_i, R_i \rangle \tag{2.12}$$

where q_i is the current status, L_i and R_i are the sequences of symbols on the tape to the left and right, respectively, of the current square x_j:

$$L_i \overset{\text{def}}{=} \langle x_1, x_2, \dots, x_{j-1} \rangle, \tag{2.13}$$

$$R_i \overset{\text{def}}{=} \langle x_j, x_{j+1}, \dots, x_n \rangle. \tag{2.14}$$

The machine starts its computation from the initial configuration:

$$C_o = \langle q_o, L_o, R_o \rangle \tag{2.15}$$

and, after a number of steps, may reach a final configuration:

$$C_f = \langle q_f, L_f, R_f \rangle. \tag{2.16}$$

Let us leave formalism aside for now, and give an example of a Turing machine. Let us have the tape populated by two possible symbols: dog and cat, with

Table 2.4: Example of a *circle-free* Turing machine that computes the infinite sequence 0101....

STATE	INPUT	NEXT STATE	CELL	OUTPUT
0	–	1	⇒	0
1	–	0	⇒	1

a third additional symbol # used to mark the left and right boundaries on the tape where cats and dogs are. We want the machine to keep dogs and cats apart, a commendable task since, according to common perception, dogs and cats fight. Since the transition function δ has the current state and the input symbol as domain, and the next state, tape cell change and output symbol as codomain, it can be written in tabular form with a transition rule – call it machine instruction – in each row. The table, with all instructions, is shown in Figure 2.3 on the left. For simplicity, we used the symbol '–' in the output column when the rule does not modify the content of the current cell, just a shortcut for the output symbol being the same as the input symbol. We arbitrarily choose to have, at the end of the execution, all dogs on the left and all cats on the right. While this machine can separate cats and dogs in any number, we applied it to an example of tape with just four pets. The extreme right of Figure 2.3 displays every step taken by this Turing machine with the position of the current cell in the tape, the changes of the symbols, and the rightmost number indicating the state of the machine. State #4 is the final state, and when the machine has reached it, we can see that the tape is well organized, with all dogs on the left and the only cat on the right.

We cannot end our description of the Turing machine without revealing what *computable numbers* are. For this purpose, it is necessary to note that there can be machines that, unlike our pet related example, never end, continuing in an infinite loop of instructions. During this infinite loop the machine can continue to write on the tape, like in the example provided in Figure 2.4. This example was given by Turing in his article, and writes on the tape the sequence 01010101010101 · · ·
The *computable numbers* are real numbers which decimal part can be written, in binary code, by a Turing machine. Turing demonstrated that there exist machines able to write many notable real numbers, like π, the real parts of all algebraic numbers, and the real parts of the zeros of the Bessel functions. There are infinite different decimals for these numbers, therefore a machine that computes these decimals should never stop.

So far, we have seen two Turing machines, those instructions are shown in Figure 2.3 and Figure 2.4. There are infinite possible machines; some may perform purposeful computations, like keeping dogs and cats apart; some just fill the tape with nonsense symbols, like the infinite sequence 01010101010101 · · ·
Turing devised an ingenious method for associating every possible machine to a natural number; this numbering allowed him to demonstrate that there are mathematical tasks that cannot be computed by his machine. His demonstration used

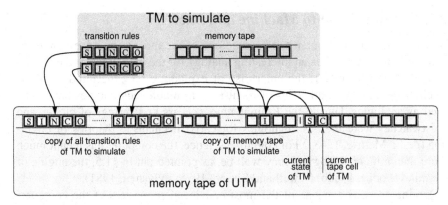

Figure 2.3: Scheme of how the universal Turing machine works, by reading on its tape all transition rules of the machine to simulate, together with its own tape. The following abbreviations are used TM = Turing machine to simulate; UTM = universal Turing machine; S = state; I = input symbol; N = new state; C = tape cell change; O = output symbol.

the *diagonalisation argument* introduced by Georg Cantor (1891) to demonstrate that, while it is possible to count all rational numbers, irrational numbers are uncountable. This result sounded the death knell of Hilbert's program: the *Entscheidungsproblem* has no solution. A sort of byproduct of the analysis done by Turing on his machine was, instead, immensely fruitful for computer science. Among the infinite possible machines, one could imagine one whose task was just to simulate the behavior of *any* other machine. This machine is feasible, and later, it came to be known as a *universal Turing machine* (Shannon, 1956).

Figure 2.3 gives an idea of how the universal Turing machine can succeed in simulating every other machine. For brevity we refer to the universal Turing machine as UTM and to the machine being simulated as TM. The UTM should read the TM as input, and this is possible since a machine is fully described by its table of transition rules. Each rule is a sequence of five symbols that belong to the domain and codomain of the transition function δ, defined in equation (2.11). In addition, the UTM should also read the input tape of the TM. Having all this information stored on its tape, the UTM can use two cells of its tape to keep trace of the current state and the current cell of the TM. These two values can be checked against each of the first two symbols in the sequence of transition rules of the TM available on the UTM's tape. When the matching sequence is found, the following three symbols instruct what to do next: change of status; cell change; output symbol. These actions can be easily performed by the UTM on the portion of tape corresponding to the original TM's tape. Today every computer is an implementation of the universal Turing machine with its own program that just simulates every possible program (software), stored in some memory like the internal hard disk.

2.2.2 From Mind to Machine and Back

By the 1960s, the computer had become one of the most impressive new technologies, and appeared as an interesting analogy for the operations of the human mind. As seen in §2.1, the enterprise of explaining how the mind works is so challenging that, at all states of its history, available technologies have been used as analogies. During the Enlightenment, gears and springs of clocks and wristwatches were the best analogies for early mechanistic theories of cognition (de La Mettrie, 1748). For cognitive science, the computer has meant much more than a mere analogy; it may well be, as pointed out in §1.5, the engine of the mind (Fodor, 1968b; Newell and Simon, 1976; Pylyshyn, 1981).

Quoting Jerry Fodor, one of the most convinced supporters of the computer analogy:

> If machines and organisms can produce behaviors of the same type and if descriptions of machine computations in terms of the rules, instructions, etc., that they employ are true descriptions of the etiology of their output, then the principle that licenses inferences from like effects to like causes must license us to infer that the tacit knowledge of organisms is represented by the programs of the machines that simulate their behavior.
>
> (Fodor, 1968a, p.640)

What is not so expected is that some thinkers do not just assume the generic computer as analogy, but adopt specifically the Turing machine as a close model of the mind, so close as to risk the Promethean frenzy that we were warned of in §1.6. Putnam (1960) introduced the Turing machine as a model of the mind in addressing the philosophical puzzle of internal state knowledge. One problem is that while it makes sense to ask how somebody is able to understand that somebody else has pain, it is odd to ask how I know I have pain. A connected question is whether mental events about ourselves coincide with physical events of our body. He proposed that these questions could be mirrored in the case of machines that become tools for better investigating these issues:

> Then we can perfectly well imagine a Turing machine which generates theories, tests them [...] In particular, if the machine has electronic 'sense organs' which enable it to 'scan' itself while it is in operation, it may formulate theories concerning its own structure and subject them to test.
>
> (Putnam, 1960, p.139)

Putnam pushed further the analogy between the Turing machine and the mind, in order to address the relationship between mental and physical states. He referred to the conventional states of a Turing machine as *logical states*, and suggested that a physically realized Turing machine has a number of additional states he called *structural states*. In keeping with the technology of his period, Putnam assumed possible structural states such as the switching of a vacuum tube. This machine may have no way of ascertaining its own structural state just as a human would; however, it can have electric sensors that copy the structural states of the vacuum tubes on the tape, so that they become available for logical states.

Putnam (1967a,b) elaborated further on the analogy of Turing machines. He extended the possibility of machines being equipped with sense organs, in sensing both the internal body and the external environment. Such machines can *prefer*, in the sense that they can assign weights to possible alternative actions, based on considerations about the consequences of such choices. Putnam was convinced strongly enough of the affinity between the mind and the Turing machine as to make the following claims:

> A Turing Machine might very well be a biological organism. The question whether an actual human being is a Turing Machine [...] or whether the brain of a human being is a Turing Machine, is an empirical question. Today we know nothing strictly incompatible with the hypothesis that you and I are one and all Turing Machines [...]
>
> (Putnam, 1967a, p.298)

The extent to which this machine is amenable to being interpreted as a mind comes naturally from Turing's inspiration in conceiving it. First of all, we should keep in mind that computers did not exist in 1936. At that time, the term 'computer' referred to humans performing mathematical calculations, since the 17th century, hired especially for computing future movements of astronomical objects in order to create celestial tables for almanacs (Grier, 2005). Turing used the analogy with humans in developing its abstract machine extensively, as highlighted by Gandy and Yates (2001, p.11–12). For example, the limitation of the machine to change tape position by one cell only, to the left or to the right, derives by the limitation of the human mental apparatus in catching more contiguous cells at a glance. Here, below, are collected passages of Turing clarifying how he derived his machine from the (human) computer:

> Computing is normally done by writing certain symbols on paper. We may suppose this paper is divided into squares like a child's arithmetic book. [...] The behavior of the computer at any moment is determined by the symbols which he is observing, and his "state of mind" at that moment. [...] Let us imagine the operations performed by the computer to be split up into "simple operations" [...] We may suppose that in a simple operation not more than one symbol is altered. [...] Besides these changes of symbols, the simple operations must include changes of distribution of observed squares.
>
> [...]
>
> We may now construct a machine to do the work of this computer.
>
> (Turing, 1936, p.298)

So, Turing proceeded in the direction opposite to that of later cognitive science: from the human mind to the abstract machine, trying to imitate the operations performed by (human) computers.

Later on, Putnam, with his typical open-mindedness, changed drastically his belief about the strict parallelism between the Turing machine and the human mind:

> In previous papers, I have argued for the hypothesis that (1) a whole human being is a Turing machine, and (2) that psychological states of a human being are Turing machine states or disjunction of Turing machine states. In this section I want to argue that this point of view was essentially wrong, and that I was too much in the grip of the reductionist outlook.
>
> (Putnam, 1975, p.298)

But cognitive science had already become well acquainted with Turing machines, and most cognitive scientists assumed that the Turing machine provides a useful model of the mind. For example Fodor, who championed the computational theory of the mind, did not endorse a strong adoption of the Turing machine as did Putnam, yet recognized it as one of the best ideas of computation:

> To a first approximation, we may thus construe mental operations as pretty directly analogous to those of a Turing machine. There is, for example, a working memory (corresponding to a tape) and there are capacities for scanning and altering the contents of the memory (corresponding to the operations of reading and writing on the tape). [...] I'm not endorsing this model, but simply presenting it as a natural extension of the computational picture of the mind.
>
> (Fodor, 1980, p.65)

Still today, there are cognitivists like Wells (2004), who reaffirm the value of the Turing machine as a model of the mind, under novel perspectives close to situated or embedded cognition.

2.3   The Imitation Game

Society games flourished in Britain during the Victorian era, especially those invented to offer indoor amusements during leisure time. These games were usually played in the best room of Victorian houses, which was the parlor. Therefore, games such as *charades and lookabout* are collected under the term *Victorian parlor games* (Beaver, 1974). The 'imitation game' invented by Turing is a reimagined Victorian parlor game; here is his description:

> It is played with three people, a man (A), a woman (B), and an interrogator (C) who may be of either sex. The interrogator stays in a room apart front the other two. The object of the game for the interrogator is to determine which of the

other two is the man and which is the woman. He knows them by labels X and
Y, and at the end of the game he says either "X is A and Y is B" or "X is B and
Y is A." The interrogator is allowed to put questions to A and B thus:
C: Will X please tell me the length of his or her hair?
Now suppose X is actually A, then A must answer. It is A's object in the game
to try and cause C to make the wrong identification. His answer might therefore
be:
"My hair is shingled, and the longest strands are about nine inches long." In
order that tones of voice may not help the interrogator the answers should be
written, or better still, typewritten.

(Turing, 1950, p.433)

Many Victorian parlor games are based on conversations from which something hidden should be guessed. One of the most popular was *Twenty questions*,
in which one player thinks of a person or a place, and the others try to guess
what he thinks of by asking questions, no more than twenty. In *How? Why?
Where? When?* there is again one player who thinks of the name of an object
that the other players try to discover by asking only questions of the format
"How/Why/Where/When do you like it?" Turing's imitation game has nothing
to envy the most popular Victorian parlor games, it looks attractive, it is rather
flirty, and it is fun. We confess that our historical curiosity has pushed us in
searching hard for evidence of people who have actually played this game. So
far, our curiosity has remained unsatisfied.

2.3.1 Can Machines Think?

The purpose of Turing was certainly not to amuse people during their leisure
time with a new parlor game as is borne out in the continuation of his paper:

We now ask the question, "What will happen when a machine takes the part of
A in this game?" Will the interrogator decide wrongly as often when the game
is played like this as he does when the game is played between a man and a
woman? These questions replace our original, "Can machines think?"

(Turing, 1950, p.434)

The question, "Can machines think?" is clearly stated at the beginning of the
paper, and ushered in the adventure towards the future artificial mind, outlined
in §1.4. Turing observed that terms like 'think' or 'intelligence' are problematic,
especially when applied to unusual domains like machines; therefore, the question as such is 'too meaningless to deserve discussion'. Turing came up with
a brilliant ploy, turning sterile debates and haggling over definitions of 'intelligence' into such a simple way to test machine intelligence. If a machine can
maintain a conversation so well that it can fool a discerning judge in this game,

Figure 2.4: Sketches of Turing's imitation games. On the left is the parlor game version, where the interrogator converses with a man and a woman, and has to guess who the woman is. On the right is the version meant to replace the question "Can machines think?" Now, the interrogator has to guess whether the agent engaged in the conversation is a human or a machine.

then this machine would be intelligent. Figure 2.4 displays the transition of the imitation game from a parlor game of sorts to a serious stratagem. It was natural for Turing to take the issue of machine intelligence seriously, while delivering a dose of British humor. For example, in an interview with *The Times* (1949, June 11), he said that he did not exclude the possibility that a machine might produce a sonnet, though it might require another machine to appreciate it. His friend, Robin Gandy (1996, p.125), reported that "He [Turing] wrote this paper unlike his mathematical papers quickly and with enjoyment. I can remember him reading aloud to me some of the passages always with a smile, sometimes with a giggle."

Nobel laureate Percy Bridgman (1950) published a paper in the same year as that of Turing, arguing that the account of meaning useful for physical scientists should be different from that usually found in linguistics, and should rely on *operations*. His *operational* account of meaning, already expounded upon (Bridgman, 1927), prescribes that terms used in scientific theories, physics in the first place, should be defined by specifying *operations* in which the term is decidedly involved. Operationalism has been influential not only in physics (Moyer, 1991), but across several scientific fields, psychology included (Feest, 2005). What has come to be known as the Turing Test can be seen as a sort of operational definition for a crucial term in psychology: intelligence. The operational definition is, at the same time, simple but general enough to be applied to artificial agents.

Since Turing's paper is one of the most highly cited papers ever written, no doubt that there is a multitude of interpretations for every aspect of his test. There are those who reject understanding the Turing test as an operational definition. The first one has been James Moor (1976):

> [...] believe that the significance of the Turing test is that it provides one good
> format for gathering inductive evidence such that if the Turing test was passed,
> then one would certainly have very adequate grounds for inductively inferring

that the computer could think on the level of a normal, living, adult human being.

(Moor, 1976, p.251)

In addition, Copeland (2000); Copeland and Proudfoot (2004) reject the operational definition view of the Turing test, but on the basis of several bibliographic cues in Turing's original aims. Still, the operational definition view is the prevailing interpretation (McCarthy and Shannon, 1956; Block, 1981b; Dennett, 1985; Haugeland, 1985; French, 1990, 2000b; Harnad, 2000).

It is interesting, the way Harnad (2000) eludes the dispute about the original view of Turing:

> Wimsatt (1954) originally proposed the notion of the "Intentional Fallacy" with poetry in mind, but his intended message applies equally to all forms of intended interpretation. It is a fallacy, according to Wimsatt, to see or seek the meaning of a poem exclusively or even primarily in the intentions of its author: the author figures causally in the text he created, to be sure, but his intended meaning is not the sole or final arbiter of the text's meaning. [...]
>
> This paper is accordingly not about what Turing may or may not have actually thought or intended. It is about the implications of his paper for empirical research on minds and machines.

(Harnad, 2000, p.253)

We wouldn't go so far as to say that seeking the genuine intent of Turing is a fallacy, but we fully agree that the enormous implications of the Turing test in its standard interpretation are as important as the hermeneutics of Turing's prose.

Turing not only spelled out the very cornerstone of AI – demonstrating intelligent behavior of machines – he also addressed a list of objections, many of which will continue to be used by future opponents of AI. The list is below, quoting Turing's formulation (pp.443–454):

1. *The Theological Objection.* Thinking is a function of man's immortal soul. [...] Hence no animal or machine can think.

2. *The 'Heads in the Sand' Objection.* "The consequences of machines thinking would be too dreadful. Let us hope and believe that they cannot do so."

3. *The Mathematical Objection.* There are a number of results of mathematical logic which can be used to show that there are limitations to the powers of discrete-state machines.

4. *The Argument from Consciousness* [...] "No mechanism could feel (and not merely artificially signal, an easy contrivance) pleasure at its successes, grief when its valves fuse, be warmed by flattery, be made miserable by its mistakes, be charmed by sex, [...]"

5. *Arguments from Various Disabilities* [...] "I grant you that you can make machines do all the one to do X." Numerous features X are suggested in this connection.

6. *Lady Lovelace's Objection.* [...] "The Analytical Engine has no pretensions to *originate* anything. It can do *whatever we know how to order it* to perform."

7. *Argument from Continuity in the Nervous System.* The nervous system is certainly not a discrete-state machine. [...] It may be argued that, this being so, one cannot expect to be able to mimic the behavior of the nervous system with a discrete-state system.

8. *The Argument from Informality of Behaviour.* It is not possible to produce a set of rules purporting to describe what a man should do in every conceivable set of circumstances.

9. *The Argument from Extra-Sensory Perception.* [...] telepathy, clairvoyance, precognition and psycho-kinesis. These disturbing phenomena seem to deny all our usual scientific ideas.

Objections 1., 2. and 9. pertain to the social and cultural acceptably of intelligent machines, and the first two continue to surface even today. All the other objections have become the basis for many of the standard criticisms of AI. There is just one objection that gives credit to previous reflections about thinking machine: 6., *Lady Lovelace's Objection*, addressing the note of Ada Lovelace we quoted in §2.1.2. Before addressing how this objection is no trouble for his test, Turing argues that Ada Lovelace did not necessarily think that no machine can originate anything, simply that the evidence available about the Analytical Engine did not encourage her to believe that it did it. In philosophical discussions about AI, the most commonly occurring argument is 4., the argument from consciousness, as we will see in §2.3.3. Turing articulated powerful rebuttals for all objections, without missing a fair amount of irony.

When Turing's paper was published there was no cognitive science nor AI, and the paper did not draw immediate attention. Pinsky (1951) quickly dismissed the Turing test, by using sarcasm, proposing to use a digital computer to read Turing's article and because "it is thinking about the possibility of machines thinking! [...] the machine suffers a nervous breakdown." A more serious and articulated criticism was raised by Mays (1951). One of his arguments was a misplaced prediction about the future of computers, which, nowadays, inevitably makes one smile:

> [...] there is an optimum limit to the size of computing machines. [...] computing machine (which is going to imitate the brain) as large as the Empire State Building and powered by the Niagara Falls, will still remain a subject for conjecture in those journals devoted to astounding science fiction.
>
> (Mays, 1951, p.154)

More interesting and lasting was Mays' fear about the perspective of intelligent machines; its echoes are still heard today:

> It is not altogether too fanciful to suppose that the machine analogy, with its emphasis on overt behavior and abnegation of private experience may [...] lead us to be regarded, more than ever before, as if we were mechanical objects.
>
> (Mays, 1951, p.154)

One of the first to take Turing more seriously was Gunderson (1964) who addressed the core question of AI: the possibility of developing intelligent machines. His answer was negative, skeptical about the provision of full intelligence to machines, although he left open the possibility for machines to have reasonable linguistic abilities. But it is only during the development of AI and cognitive science that the imitation game became one of the most iconic and controversial aspects of AI, bringing about a mass of discussions about Turing and his ideas. Turing's paper included a confident prediction in the possibility of computers able to sustain his test 'in about fifty years' time', this timing has given rise to a huge resurgence of publications after the fateful date of 2000 (Saygin et al., 2000; French, 2000b; Moor, 2003; Shieber, 2004; Teuscher, 2004; Epstein et al., 2008; Warwick and Shah, 2016; Levesque, 2017). A further boost to studying and spreading Turing's ideas was the centenary of his birth in 2012, celebrated as the Alan Turing Year. The outcome of the celebrations that mostly spread the fame of imitation game was, as mentioned in §2.2, the eponymous film by Morten Tyldum. In fact Turing had already been the protagonist of literary works, like the plays *Breaking the Code* by Hugh Whitemore (1986), and *Alan's Apple: Hacking the Turing Test* by Valeria Patera (2004).

We will delve more deeply into the philosophical debate triggered by the Turing test in §2.3.3, following the birth of AI. For now, let us notice some unexpected or extravagant reflections and elaborations on the Turing test. Genova (1994) is the most representative of the few scholars who regard gender as an important aspect of the imitation game, and suggests that Turing's inhibition in procreating another intelligent being may have pushed him into conceiving alternative ways of creating thinking entities. Collins and Evans (2014) borrowed Turing's imitation game as a method in social sciences for investigating the distribution of interaction expertise within and between societies. In their own words:

> It [the method] is a more rigorous version of the parlor game that inspired Alan Turing's famous "Turing Test" for the intelligence of computers. [...]

> One participant, drawn from the "target group" whose abilities are being investigated acts as the judge. The judge creates questions and sends these to the other two participants. Of these, one is another member of the target group and is asked to answer the questions naturally. The other is drawn from a different group and charged with answering as if they were a member of the target group. The judge then compares the answers and tries to work out which comes from the person who is pretending and which from the person who is answering naturally.
>
> (Collins and Evans, 2014, p.7)

Ward (2018) draws a parallel between the human verisimilitude of machines, and aspects such as mechanicity and repetitive domesticity in some characters of Victorian novels. Turing's imitation game is used by Ward to read the flat characters Lizzie Eustace and Ferdinand Lopez in *The Prime Minister* by Anthony Trollope (1876). The videogame *Detroit – Become Human*, released in 2019 by Quantic Dream, features Elijah Kamski as the scientist who develops androids able to pass the Turing test. He invented a more demanding test, the Kamski Test (Tomczak, 2019), where an android is offered a deal. He could another android in exchange for obtaining vital job information. If the android refrained from killing, it then demonstrated empathy, the last remaining feature that can distinguish humans from androids.

2.3.2 Russell Demonstrated by Machine

The first functioning computer program demonstrating some form of intelligence did not converse and, therefore, was not a candidate for the Turing test. It rather followed the tradition from which the Turing machine was born, as addressed in §2.1.1: logic. The program named *Logic Theory Machine* – often called *Logic Theorist* – was developed by Newell et al. (1957). The ambition of the program was to grasp some of the peaks of human thought, demonstrating theorems of *Principia Mathematica* (Whitehead and Russell, 1913), the bible of logic at that time. The Logic Theorist succeeded in demonstrating thirty-eight of the theorems, starting from the five axioms and three rules of inference in *Principia Mathematica*. It even proved one of the theorems with a shorter and more elegant solution than that given by the original authors; Simon conveyed this news to Russell who responded with delight (McCorduck, 1979, p.167). The possibility of a computer demonstrating theorems of *Principia Mathematica* was already conjectured ten years earlier by Turing (1948a), who didn't attempt to write such a program.

The Logic Theorist starred at the 1956 Summer Research Project held at Dartmouth College, New Hampshire, widely regarded as the kick–off point for AI. The event was planned by McCarthy, Minsky, Rochester and Shannon being the only one who was already famous at that time. The incipit of the proposal was the following:

We propose that a 2 month, 10 man study of artificial intelligence be carried out during the summer of 1956 at Dartmouth College in Hanover, New Hampshire. The study is to proceed on the basis of the conjecture that every aspect of learning or any other feature of intelligence can in principle be so precisely described that a machine can be made to simulate it.

(McCarthy et al., 1955, p.12)

Actually Newell and Simon had preferred the term *complex information processing* for the new field, but McCarthy had insisted on using 'artificial intelligence', a farsighted choice. The Logic Theorist was presented with enthusiasm at Dartmouth as the first working AI program, even if it is historically arguable (Boden, 2008, p.705).

Ten years later, the first wave of artificial conversational programs, *Eliza*, was launched by Joseph Weizenbaum (1966). If we were to find an AI program that is poles apart from the austere and rigorous mathematical demonstrations of the Logic Theorist, it is Eliza. Despite being capable of an only extremely rudimental conversation, it become – probably – the most widely distributed program in the history of AI (Wallace, 2008, p.186). Eliza is amusing, and slightly irreverent, because it is the parody of a Rogerian psychotherapist. The choice of the counseling practice founded by Carl Rogers (1942) is smart, because the practitioner should always appear sympathetic with the patient, asking only bland questions, often asking the patient to clarify what he has just said. Therefore, the therapist does not have to engage in any detail with the real problem of the patient, and Eliza can happily ignore all references to the real world in the patient's sentences. Here is a short excerpt from one of the many current online interfaces to Eliza[3], with one of us as the patient:

patient	I think I need psychological help
Eliza	Do you really think so?
patient	I have relationship problems with women
Eliza	Say, do you have any psychological problems?
patient	this is my main problem: I am afraid of women
Eliza	Did you come to me because you are afraid of women?
patient	yes, I hope you can help me
Eliza	We were discussing you, not me.
patient	How can I get out of this problem?
Eliza	Do you want to be able to get out of this problem?

In fact Weizenbaum did not state any intention to deride psychotherapy; on the contrary, his motivation for choosing this domain of conversation would seem sincerely benevolent:

---

[3] http://psych.fullerton.edu/mbirnbaum/psych101/Eliza.htm.

> [...] ELIZA to respond roughly as would certain psychotherapists (Rogerians). ELIZA performs best when its human correspondent is initially instructed to talk to it [...] just as one would to a psychiatrist. This mode of conversation was chosen because the psychiatric interview is one of the few examples of categorized dyadic natural language communication in which one of the participating pair is free to assume the pose of knowing almost nothing of the real world. [...] Whether it is realistic or not is an altogether separate question. In any case, it has a crucial psychological utility in that it serves the speaker to maintain his sense of being heard and understood.
>
> (Weizenbaum, 1966, p.42)

Eliza was actually amazingly able to instill this "sense of being heard and understood", far beyond Weizenbaum's expectations. Anecdotes tell that Weizenbaum's secretary felt so comforted by the conversation with Eliza as to ask Weizenbaum to leave the room in order to talk to the machine privately (Block, 1981b, p.233). Despite the success of Eliza in the eyes of the general public, the young AI followed the path of the Logic Theorist instead, as we will describe in §3.1. Eliza was viewed with suspicion, and even a certain contempt, for achieving so much with so little inside. It was just a simple pattern matcher, with a number of predefined responses triggered by keywords or phrases, and a set of rules for transforming part of the question into the answer, triggered by certain keywords. After Eliza, the Turing test was not a priority for AI and, in fact, no one had thought to set up something that resembled a formally regulated Turing test. No one until Hugh Loebner, an eccentric American inventor and entrepreneur, established in 1990 the Loebner Prize, offering a $10,000 prize for the first programmer who could pass the test. We will see Eliza's legacy contending at the Loebner Prize in §3.2.

2.3.3 Into a Chinese Room

As long as AI and cognitive science grow, the Turing test will remain as the flagship of the new movement, and will be widely discussed, attacked, and defended over and over. One of the most prominent philosophers in support of the imitation game is Daniel Dennett (1985), who even headed the scientific committee of the first Loebner Prize Contest in 1991. There is a strong convergence between the account of intelligence as an observable attribute rather than an intrinsic one in the Turing test, and Dennett's view of mentalistic notions as attributed properties. A few years later, Dennett (1987) would elaborate on this thesis under the famous label of 'intentional stance'.

Dennett contrasted his admiration for Turing's idea with his bewilderment for the difficulties of other philosophers in receiving such simple and clean idea:

> It is a sad irony that Turing's proposal has had exactly the opposite effect on the discussion of that which he intended. Turing didn't design the test as a useful

tool in scientific psychology, a method of confirming or disconfirming scientific theories or evaluating particular models of mental function; he designed it to be nothing more than a philosophical conversation-stopper. He proposed [...] a simple test for thinking that was *surely* strong enough to satisfy the sternest skeptic [...] Alas, philosophers – amateur and professional – have instead taken Turing's proposal as the pretext for just the sort of definitional haggling and interminable arguing about imaginary counterexamples he was hoping to squelch.

(Dennett, 1985, p.2)

The list of defenders of the Turing test is long (Moor, 1976; French, 2000b; Rapaport, 2000; Hauser, 2001), albeit few are as sharp as Dennett. Let us now look at those philosophers accused by Dennett of 'interminable arguing' about Turing's proposal, leaving the 'amateur' ones out. Ned Block's (1981a) first charge against the Turing test is of behaviorism was repudiated with shame by the cognitive sciences of that period. Block offers a way out of this shame by moving the requirement for a machine to actually display its behavior during the test, to just having the *capacity* to pass the test. This modification saves the test from standard arguments against behaviorism, but Block invented a mental counterexample supposed to defeat the test anyway. He supposed a program where all possible questions are stored as symbols, with a corresponding short list of possible questions, possibly inspired by the answer the aunt of the programmer, Bertha, might give to the same questions. This sort of machine emulates the intellectual behavior of a human (aunt Bertha), but does not think. In addition to this main argument, Block took Eliza as a wicked specimen of an AI program that pretends to be intelligent, and may even pass the Turing test:

Though this system [Eliza] is *totally* without intelligence, it proves *remarkably* good at fooling people in short conversation. [...] the program's extraordinary success [...] reminds us that human gullibility being what it is, some more complex (but nonetheless unintelligent) program may be able to fool most any human judge.

(Block, 1981a, p.233)

Block was unlucky, because almost in coincidence with his writing, John Searle (1980) invented a very similar mental experiment, bound for lasting fame. To anyone who has even heard of AI or cognitive science the 'Chinese room' – depicted as a cartoon in Figure 2.5 – is something familiar, not so much 'aunt Bertha'.

Suppose that I'm locked in a room and given a large batch of Chinese writing. [...] Now suppose further that after this first batch of Chinese writing I am given a second batch of Chinese script together with a set of rules for correlating the second batch with the first batch.[...] Now suppose also that I am given

Figure 2.5: Cartoon portrait of John Searle in his Chinese room.

a third batch of Chinese symbols together with some instructions, [...] that enable me to correlate elements of this third batch with the first two batches, and these rules instruct me how to give back certain Chinese symbols with certain sorts of shapes in response to certain sorts of shapes given me in the third batch. [...] Suppose also that after a while I get so good at following the instructions for manipulating the Chinese symbols [...] that is, from the point of view of somebody outside the room in which I am locked – my answers to the questions are absolutely indistinguishable from those of native Chinese speakers. [...]
it seems to me quite obvious in the example that I do not understand a word of the Chinese stories.

(Searle, 1980, p.418)

Probably everybody who read this story for the first time was persuaded that AI is wrong. No, machines can well pass the Turing test, but they will never understand, they will never think! The persuasive power of this story has mercilessly overshadowed Block's 'aunt Bertha' thought experiment, despite their strong similarities. Moreover, Searle has broadened his attack on the Turing test to include the whole enterprise of AI developing computers that 'think'. Searle's Chinese Room experiment led Dennett to invent the term 'intuition pump', which was to be the subject of a later book with the same title (Dennett, 2012). Here, in the commentary to Searle, is what Dennett meant by 'intuition pump':

a device for provoking a family of intuitions by producing variations on a basic thought experiment. An intuition pump is not, typically, an engine of discovery, but a persuader or pedagogical tool – a way of getting people to see things your way once you've seen the truth, as Searle thinks he has. [...] Searle relies almost entirely on ill-gotten gains: favorable intuitions generated by misleadingly presented thought experiments.

(Dennett, 1980, p.430)

This 'intuition pump' has tremendous appeal and has been debated over and over for more than twenty years, with entire books devoted to its debate (Preston and Bishop, 2002). The enchantment power of this short story amazed Dennett (2012, 192): "We [Dennett and Hofstadter] found his thought experiment fascinating because it was, on the one hand, so clearly a fallacious and misleading argument, yet, on the other hand, just as clearly a tremendous crowd-pleaser and persuader. How – and why – did it work?".

We leave Dennett to scrutinize the subtle strategies that make Searle's 'intuition pump' so convincing; here we point to just a couple of 'fallacious and misleading' aspects of the argument. The main one is the imagined large batches of Chinese symbols that allow any non Chinese speaker to carry on an impressive conversation in Chinese. Of course these batches should be continuously updated, so that it would be possible to converse about the latest television news. Such a database is literally impossible, there are infinite ways to compose questions and answers in conversations. Nevertheless, let us concede that this database exists, and is manageable. Then, the individual person who is locked in the room has a negligible role; it is the whole system, at the heart of which there is the fabulous database, that enable the conversation. There is no reason to deny that the whole system understands the ongoing conversation.

Dennett was certainly not the only one to debunk Searle's thought experiment (Hofstadter, 1981; Harnad, 1989; French, 2000a), that, today, has lost its mooring. Still, in contemporary attacks on current AI it is often possible to find conceived traces of the Chinese room, as we will see in 5.3.3.

Chapter 3

The Early Shaping of Cognitive Science by Artificial Intelligence

In the previous chapter, we saw how the amazing *Logic Theory Machine* by Simon and Newell celebrated the baptism of AI in 1956. The same year, program was the main event of another conference, the Symposium on Information Theory held at the MIT, considered the event from which cognitive science sprang (Gardner, 1985, p.28). It is certainly not a coincidence; we can say that the ideas of Newell and, even more, of Simon, constitute the closest link between the first AI and early cognitive science. Simon's theory of bounded rationality – that garnered him the Nobel Prize – was at the same time, the structure on which artificial models of rational behavior could be developed, and the interpretative key to explain cognitive behaviors. This is the topic of the first section of this chapter. The same day that the *Logic Theory Machine* was presented at the Symposium on Information Theory, the paper *Three Models of Language* by Noam Chomsky was also featured. It is difficult to determine who, between Chomsky and Simon, most influenced cognitive science in its early days. However, in contrast to Simon and Newell, Chomsky has maintained a strong distance from AI, despite the extensive application of formal grammar in computer science, and in AI. In the symbiotic progress of cognitive science and AI in the 1960s, the field of perception was left out. It was David Marr who brought clarity to this area as well, with the separation into levels of description of a cognitive system, which also made possible the assumption of symbolic information processing for visual perception.

3.1 Solving Problems

Herbert Simon and Allan Newell brought together a perfect idyll between AI and cognitive science, an idyll that would never be experienced again. Rarely has an AI program been so deeply inspired by human cognition than those developed by Simon and Newell. Conversely, few ideas of AI programs have been as influential in setting the cognitive science approach as Newell and Simon's work in mid-1950s. Actually a program like the *Logic Theory Machine* (see §2.3.2) would not seem the best archetype for cognition; in the end, very few people spend their time thinking about Russell's theorems and how to solve them. But Simon's intent was really to simulate, in a computer program, the way humans reason. The choice of Russell's theorems was because, during that period, it was a kind of bible of logic, and because engaging in their solution certainly requires intelligence. In fact Russell was a fallback; initially, the goal of Simon was supposed to be the more traditional AI one of playing chess, but he realized that vision plays an important role in chess, and it was too difficult to simulate. An abstract task, independent from perception, looked easier to simulate.

Simon made his entrance into AI late, after a brilliant career as economist. His early expertise was in administrative organizations (Simon, 1947), but his best known achievement is the concept of *bounded rationality* in economy (Simon, 1957), that garnered him the Nobel prize. Bounded rationality is a fundamental evolution over the *expected utility theory* (Bernoulli, 1738), according to which rational agents ought to maximize expected utility. The crucial drawback of expected utility theory is that it entails a complete knowledge of all that logically follows from one's choice, Simon proposed, instead, to replace this view with the more plausible rational behavior that is compatible with the access to cognitive capacities, which are necessarily bounded.

Here is how Simon, in his autobiography, narrates his sudden entrance into AI:

> During the preceding twenty years, my principal research had dealt with organizations and how the people who manage them make decisions. [...] My theorizing used ordinary language or the sorts of mathematics then commonly employed in economics. [...] All of this changed radically in the last months of 1955. While I did not immediately drop all of my concerns with administration and economics, the focus of my attention and efforts turned sharply to the psychology of human problem solving, specifically, to discovering the symbolic processes that people use in thinking. Henceforth, I studied these processes in the psychological laboratory and wrote my theories in the peculiar formal languages that are used to program computers. Soon I was transformed professionally into a cognitive psychologist and computer scientist, almost abandoning my earlier professional identity.

> (Simon, 1991, p.204)

Simon's quick success in AI was favored by collaborations formed during his period at RAND Corporation in Santa Monica as consultant for the human organization within air-defense military systems. There, he met Allan Newell, a mathematician working on logistics problems of the Air Force, and Cliff Shaw, a systems programmer. A detailed historical account of their encounter at RAND Corporation is in (McCorduck, 1979, Ch. 6). Simon's intellectual roots for the development of the *Logic Theory Machine* were profuse. His view on administrative organization involved decision making under bounded cognition, the use of heuristics, the definition and achievement of sub-goals relevant to a main task.

Galvanized by the success of the *Logic Theory Machine*, Simon and his collaborators abandoned the exclusive world of Russell's theorems and ventured into the wild with explorations of cognition. The new system sought to simulate the mental processes involved when solving any kind of problem; it therefore was called GPS (*General Problem Solver*) (Newell et al., 1959; Newell and Simon, 1972).

3.1.1   *Reasoning with Bounded Cognition*

In GPS a 'problem' is basically a state of affairs that an individual want to attain, but it is not immediately clear how to do so. This desired state of affairs and the current state of the individual (and its environment) are defined as *objects*, and are represented by abstract symbols. Objects have associated features, and GPS includes routines for measuring the *difference* between two objects, based on their features. Solving a problem is the transforming of objects until there is no difference between the current state and the desired state. Transformations are done by applying *operators* to objects; the choice of the appropriate operators, and their order, is planned by heuristics. This concept has been the workhorse of Simon, right from his research on decision making in administrative organizations (Simon, 1947) to the process of scientific discovery (Simon, 1977). Heuristics is intimately related with bounded rationality; it deals with partial available information and limitation in the number of possible steps to evaluate. It is, therefore, a sub-optimal practical guidance. It may suggest the order in which possible solutions should be examined; it may rule out classes of steps; it may provide a quick screening to select the more promising steps.

The basic heuristics in GPS derives from *means–ends analysis*, a general method common in social and economic sciences (Weber, 1921; Robbins, 1932) and in policy science (Hyneman, 1959). By taking into account possible means and ends, GPS has the following three *goal types*:

1. Find a way to transform object a into object b;

2. Apply operator q to object a or to an object obtained from a;

3. Reduce the difference, d, between object a and object b by modifying a.

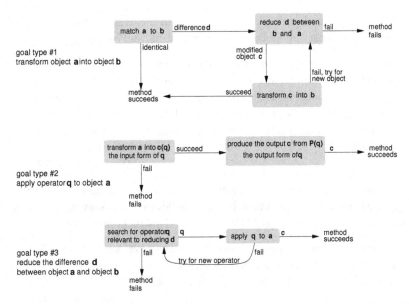

Figure 3.1: Methods used in the three goal types of GPS, from (Newell et al., 1959, p.10).

For each of the three goal types, GPS has corresponding methods, shown in Figure 3.1. Note that methods may in turn set new subgoals; for example, the method associated with goal type #1 (top of Figure 3.1) consists of measuring the difference d between objects a and b, and consequently setting up the goal type #3 of reducing d; if the method (bottom of the figure) succeeds with output c, a new subgoal of type #1 is set up of transforming c into b. All subgoals – and sub-subgoals – are set up as means to attaining the end. If a chosen step could not be made to work, GPS can backtrack to the previous choice point and try an alternative step. There is no guarantee of success, and this is natural in the realm of bounded rationality. GPS was truly 'general' in the sense that the specific content of the problem to solve was irrelevant, unlike for the *Logic Theory Machine* where problems must have the format of mathematical theorems. However, GPS requires the effort of representing the problem in terms of objects, features, differences, transformations. For real problems, this may require a great effort and the choice made in the representation of the problem would drastically affect the efficiency of the solver.

But GPS was not intended for practical usage, it was a working concept that turned out to be incalculably significant for the history of both AI and cognitive science. The central claim is that minds and computers are exemplars of the same class, that of the *physical symbol system* (Newell and Simon, 1972; Newell, 1980). According to Gardner (1985, p.163) the concept of physical symbol system has its crucial place in cognitive science "just as the cell doctrine has proved central in biology, and germ theory is pivotal in the area of disease." Nor did Simon display modesty about the significance of his and Newell's work:

> we invented a computer program capable of thinking non-numerically, and thereby solved the venerable mind/body problem, explaining how a system composed of matter can have the properties of mind. With that, we opened the way to automating a wide range of tasks that had previously required human intelligence, and we provided a new method, computer simulation, for studying thought.
>
> (Simon, 1991, p.204)

There are curious notes about how firmly Simon was convinced, on a personal level, of the mind as an information processing device. One, known as the *Travel Theorem*, is picked up by Dupuy (1994, Ch. 1). This theorem states that:

> Anything that can be learned by a normal American adult on a trip to a foreign country (of less than one year's duration) can be learned more quickly, cheaply, and easily by visiting the San Diego Public Library.
>
> (Simon, 1991, p.343)

Firmly believing his theorem, the first time Simon and his wife visited Europe, in 1965, they arranged their itinerary so that they would see nothing that they did not already know of through books.

3.2   Formal Languages and Cognition

The attribute 'formal' can have different meanings when applied to 'languages'. Within the domain of logic, Dutilh Novaes (2012) identified at least two main different notions: formal as *computable* and formal as *de-semantification*. The latter corresponds to the use of symbols and constructs abstracting from all meaning whatsoever. We met in §2.1.1 the constructions of formal languages for expressing reasoning, from Boole to Frege and Russell. The variables in these formalisms abstract from most semantic aspects of propositions, but still preserve a fundamental semantic property of linguistic expressions: their truth value. For this reason, logic is often considered a formalism that captures the *semantics* of linguistic expressions. What is definitely irrelevant in logic is the form of the words that constitute a proposition. The formality we are going to describe in this section is even more *de-semantified* that in Novaes' category. It is the formal description of sentences introduced by Noam Chomsky, that reduces to a computational form the words of a sentence, taking care of their syntactic roles only. As we will see, in fact, Chomsky expanded on a previous mathematical framework for describing combinations of words, independent not only from their lexical meaning, but from syntax also. The figure of Chomsky has been central to cognitive science since the 1960s, but his work was soon borrowed by computer science and AI, despite the indifference of Chomsky towards these disciplines.

3.2.1 The Fascination with Combinatorics on Words

Words, stripped of their meaning, and appreciated for their character composition, easily captivated mathematicians. One of the first was the great German mathematician Friedrich Gauss, who invented a code generated by a self-intersecting closed curve in the plane (Gauss, 1900, pp.282–286). The so-called *Gauss code* of a curve is obtained by labeling all intersection points with distinct symbols, then proceeding along the curve and noting each crossing point label as it is traversed. An example is shown in Figure 3.2.

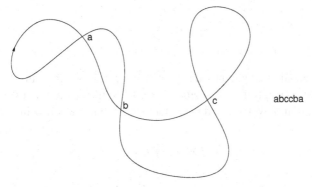

Figure 3.2: Example of *Gauss code*: a self-intersecting closed curve, with labels at the intersections; on the right is the word obtained by proceeding along the curve, starting from the arrow.

The first fundamental combinatorics on words was established by the Norwegian mathematician Axel Thue (1906, 1912). Not only had the work of Thue have little relationship with linguistics, it had no applications at all; it was just an intellectual rewarding engagement. Thue clarified where he stood on the issue of applications in the introduction of his 1912 paper:

> Für die Entwicklung der logischen Wissenschaften wird es, ohne Rücksicht auf etwaige Anwendungen, von Bedeutung sein, ausgedehnte Felder für Spekulation über schwierige Probleme zu finden.[1]

There is no doubt that the combinatorics on words is rich with difficult problems, what Thue did not foresee was the tremendous impact of his pleasant speculations on the domains of linguistics and computer science. Words, even if still called 'words', are not only meaningless, but can also be made up of symbols other than the letters of the alphabet. The symbols should be defined by a finite set, still called 'alphabet'. The basic operation on words is the *concatenation*, just putting words one after the other. Being the most common operation of words, it

---

[1] For the development of the logical sciences it will be important to find large fields for speculations about difficult problems, without consideration for possible applications.

has no specific symbol, quite like the product in algebra: if w and v are two words, their concatenation is simply written as wv. Given an alphabet of basic symbols A, by repeatedly applying concatenation to those symbols, and to the results of concatenations, infinite different words can be generated. In modern abstract algebra jargon the set of all these words is denoted by A^*, and called the Kleene star of A, after the American mathematician Stephen Kleene (1952). The set A^*, together with the concatenation operation and the empty word ε is the type of object called *monoid* in abstract algebra, that has the following properties:

1. associativity: $\forall w, v, u \in A^* (wv)u = w(vu)$;

2. closure: $\forall w, v \in A^* wv \in A^*$;

3. identity: $\forall w \in A^* w\varepsilon = w$.

Thue was especially intrigued by patterns inside words, for the study of which he introduced the concept of *factorization*, extending the analogy between product and concatenation. Given a word w, its factorization is the following:

$$w = \prod_{i=1}^{n} u_i^{k_i}, \tag{3.1}$$

where all u_i are the most basic subwords of w, their *factors*. A word is *k-free* if there is no factor of order k, and is k^+-free if there is factor of higher order than k. k. An *overlap* is a subword with factors $vuvuv$, the smallest for which v is at the same time a prefix, a suffix, and an internal factor. A word is *overlap-free* if there is no overlap inside it.

The most advanced and powerful tool introduced by Thue in analyzing word patters is the concept of *rewriting rule*, an operation that replaces a subword in a word with a different pattern. If the initial word $w \in A^*$, then the result w' of the application of a rewriting rule to w belongs to A^*. In abstract algebra, a rewriting rule is an *automorphism* with respect to the operation of concatenation, i.e., a mapping $\phi : A^* \Longrightarrow A^*$ for which it holds the following:

$$\phi(vw) = \phi(v)\phi(w), \quad v \in A^*, w \in A^* \tag{3.2}$$

We would like to give as example of rewriting rules, the most famous one proposed by Thue, based on an alphabet with just two symbols $A = \{a, b\}$:

$$
\begin{aligned}
a &\Longrightarrow ab \\
b &\Longrightarrow ba
\end{aligned}
\tag{3.3}
$$

The rewriting system with the above two rules can be applied over the result of its application, the composite operation known as *iterating morphism*. For example, starting with the word $w^{(0)} = $ a, we have the following results:

$$
\begin{aligned}
w^{(0)} &= \text{a,} \\
w^{(1)} &= \text{ab,} \\
w^{(2)} &= \text{abba,} \\
w^{(3)} &= \text{abbabaab,} \\
w^{(4)} &= \text{abbabaabbaababba.}
\end{aligned}
$$

These words are called *Thue-Morse words*, from Marston Morse (1921) who contributed to their study a few years later than Thue. Thue proved that Thue-Morse words are 2^+-*free*, they do not contain factors of order higher than two.

In rewriting systems, the rules, like those in Equation (3.3), have the following general format:

$$
\alpha x \beta \implies \alpha y \beta \tag{3.4}
$$

where α, β are variables in A^*, and x, y are fixed words in A^*. The combination of all words of an alphabet A and a rewriting systems with rules of the type (3.4) establishes a subset of A^*, and is called a *semi-Thue* system. A compelling question Thue asked for rewriting systems is whether a word w can derive from another word v by applying rewriting rules, for any $w, v \in A^*$. About half a century later, Emil Post (1947) gave the answer, demonstrating that this problem is undecidable.

Soon, we will see how the systems developed by Thue were the grounds for modern linguistics and computer programming. However, the study of meaningless words for pure intellectual pleasure has always survived, albeit in a very restricted community. After Thue, during the first half of the previous century, there were only a few isolated works; the field grew again in France with the research of the biologist and mathematician, Marcel-Paul Schützenberger (1955). Recent developments, with several results relevant for number theory, are due to Lothaire (1997, 2002)[2].

3.2.2 *Words in Ordinary and in Programming Languages*

Chomsky (1956, 1957) used the early mathematical tools for words developed by Thue and Post for an ambitious project, a fully formalized theory of natural language. Soon Chomsky clarified that, for him, a linguistic theory shall provide an account of syntax only, leaving semantics out.

The example given in Chapter 4 of *Syntactic Structure* (Chomsky, 1957) is the minimum grammar for the sentence shown in Figure 3.3.

```
the man hit the ball
```

---

[2]Note that there is no mathematician named M. Lothaire, it is a collective pseudonym derived from Lotharingia, an ancient name for the area comprising Lorraine, Alsace, Belgium, Luxembourg, from where most of the authors are. Many of this group of authors were students of Schützenberger. The use of a collective pseudonym is not unique among French mathematicians, famous is Nicolas Bourbaki.

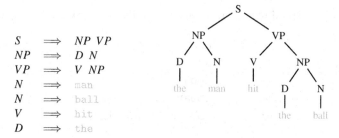

$$
\begin{aligned}
S &\implies NP\ VP \\
NP &\implies D\ N \\
VP &\implies V\ NP \\
N &\implies \text{man} \\
N &\implies \text{ball} \\
V &\implies \text{hit} \\
D &\implies \text{the}
\end{aligned}
$$

Figure 3.3: The first simple example of grammar given by Chomsky (on the left), and the corresponding syntactic tree (on the right). The terminal elements are shown in gray.

In the structure of the sentence as a tree, all nodes except leaves correspond to rewriting rules in the grammar, where the left elements are *non-terminal*. The root of the tree is the element *S*, for *Sentence*, all leaves are *terminal* elements, for which there are no rewriting rules. In practice, the terminal elements are the lexicon of a language, and the leaves of a syntactic tree, read from left to right, correspond to the given sentence. In Table 3.1 are the most common non-terminal elements of a grammar.

The project outlined in *Syntactic Structure* was pursued in the direction of mathematical rigor by Chomsky (1958); Chomsky and Miller (1963), even in collaboration with a specialist of combinatorics on word here mentioned in §3.2.1 (Chomsky and Schützenberger, 1959).

At the core of the mathematical description of language is the definition of abstract grammar:

$$
G \stackrel{\text{def}}{=} \langle V_N, V_T, S, R \rangle \tag{3.5}
$$

The elements are:

$$
\begin{array}{ll}
V_N & \text{set of non terminal elements} \\
V_T & \text{set of terminal elements, with } V_N \cap V_T = \emptyset \\
S & \text{root element, with } S \in V_N \\
& \text{set of rewriting rules, each with the format } \langle P, Q \rangle \\
R & \text{with } P, Q \in (V_N \cup V_T)^* \text{ and } P \text{ with at least on ele-} \\
& \text{ment } \in V_N
\end{array} \tag{3.6}
$$

A 'language' is now defined as follows:

$$
L(G) = \left\{ P \mid S \underset{G}{\overset{*}{\implies}} P \right\} \cap V_T^* \tag{3.7}
$$

where the operation $\underset{G}{\overset{*}{\implies}}$ is the application of any rule available in *G* an arbitrary number of times. Equation (3.7) says that a language is the set of all sentences generated according to a grammar of the vocabulary of terminal elements V_T.

Table 3.1: The most common non terminal elements, with the corresponding grammatical category.

S = *Sentence*	**N** = *Noun*
VP = *Verbal Phrase*	**V** = *Verb*
NP = *Noun Phrase*	**A** = *Adjective*
PP = *Prepositional Phrase*	**P** = *Preposition*
AP = *Adjectival Phrase*	**D** = *Determinant*

This abstract construct was just the basis for building a set of rules corresponding to the extant combination of words in the sentences of natural languages. One of the most serious challenges in this enterprise was the definition of a Universal Grammar able to explain the great variety of syntactic features in the different languages of the world. This effort has led to a lengthy evolution of the theory itself, reflected in a sequence of labels. In the Government and Binding theory (Chomsky, 1981) the focus was on the relation between two words, not necessarily adjacent in the sentence, and the binding with anaphoric elements like pronouns. It was superseded by the Principles and Parameters (Chomsky and Lasnik, 1993) theory, in which a small number of syntactic constructs, such as the X-bar, are held to be universal, with 'parameters' that adapt their precise format to each language of the world. Throughout this evolution there has been a progressive relinquishing of the stringent mathematical formalization, which has become a barely perceivable background in the most recent elaboration of the theory, the Minimalist Program (Chomsky, 1993, 1995), where the only syntactic operation left is Merge, roughly working as a rewriting rule.

Chomsky had an enormous influence on early cognitive science. In his introduction to the second edition of *Syntactic Structure*, David Lightfoot affirms that "Noam Chomsky's *Syntactic Structures* was the snowball which began the avalanche of the modern cognitive revolution." Chomsky advanced his reputation as champion of cognitive science with his harshly critical review (Chomsky, 1959) of Burrhus Skinner's *Verbal Behavior* (1957). In this very long review, Chomsky claimed that behaviorist theories are inadequate to represent grammatical structure in principle, and that behaviorist classical concepts, such as stimulus, response, and reinforcement were being applied to language by Skinner so loosely as to be vacuous. Chomsky's review attracted enormous attention not only for the compelling arguments, but also for his sparkling, stinging wit.

Pursuit of behaviorism aside, the syntax as formalized by Chomsky, made of categories of symbols and rules for their composition, was an attractive vision for what might underline the operations of the mind. In addition, the universal grammar idea was well in line with the rationalist approach of Newell and Simon in AI, seen in §3.1, intimately linked with early cognitive science. However, Chomsky and his school pushed the cognitive relevance of abstract grammar even further, with high claims on their mathematics as corresponding to mental processes of language understanding, and even brain processes. These too bold and untenable claims began to weaken the effective value of Chomskyan linguistics for cognitive science. Katz (1981) was one of the first to call attention to the issue in ontological terms. The objects of analysis in the generative grammar framework are abstract entities, mathematical objects. Yet Chomsky insisted that they are psychological entities. He did not deny the ambiguity (Chomsky and Halle, 1968): "We use the term 'grammar' with a systematic ambiguity. On the one hand, the term refers to the explicit theory constructed by the linguist and proposed as a description of the speaker's competence. On the other hand, it refers to this competence itself." The fallacy is the following:

> We may assume that there is a domain of fact, A, instantiated by (1)–(6) [syntactic rules], studied in field A' and a domain of fact, B, concerned with human linguistic knowledge, its development, the biological structures [. . .] which determine it, etc., studied in field B'. Evidently, both domains A and B and fields A' and B' are characterized a priori in distinct ways. While A and B could turn out to be identical, they could also turn out to be distinct. Therefore, they cannot simply be *assumed* to be identical.
>
> (Katz and Postal, 1991, p.527)

Similar ontological objections have been raised by Seuren (2004) and by Stich and Ravenscroft (1994). The ambiguous and contradictory ontology and epistemology in generative grammar have worsened throughout the years, as the aspirations have grown higher, from mentalism and psychologism up to the physical level of brain and biology, in what is called Chomsky's 'biolinguistic' view:

> It [the biolinguistic approach] is concerned with mental aspects of the world which stand alongside its mechanical, chemical, optical and other aspects. It undertakes to study a real object in the natural world – the brain, its states, and its functions.
>
> (Chomsky, 2000, p.5)

A longstanding consequence of the ambiguity between the ontological status of generative grammar and neurophysiological reality is the postulate of a 'language organ', often called the 'language faculty'. According to Chomsky (1995, p.167): "The human brain provides an array of capacities that enter into the use and understanding of language (the *language faculty*)". Obviously the hypothesis of a domain-specific, innate module corresponding to Chomsky's language faculty has not garnered substantial support from neurobiology (Stowe et al., 2004; Osterhout et al., 2007; Proverbio et al., 2009).

The clash with cognitive science was even more puzzling, in that Chomskian linguistics departed from psychology programmatically, in disregarding the connection between syntax and meaning. There has been, in the Chomskyan school, a group of linguists working on rejoining cognitive science, by combining syntactic structures with semantic representations. The project, named 'generative semantics', was pursued by linguists such as Postal, Lakoff, and McCawley. Chomsky was displeased and mounted a campaign against generative semantics, popularized as the 'linguistics war' (Harris, 1993), causing its rapid disappearance. In the meantime, cognitive science was growing, and gradually reacted to Chomsky with several alternative views, collected under the label of 'cognitive linguistics' (Croft and Cruse, 2004; Geeraerts, 2006; Brdar et al., 2011). Syntactocentricsm was the main issue that persuaded Lakoff (1986) to abandon the generative grammar enterprise, becoming one of the leading exponents of cognitive linguistics, gradually followed by several others (Langacker, 1987; Goldberg, 2006; Tomasello, 2005).

While the abstract grammar developed by Chomsky turned out inadequate for the cognitive faculty of language, it has become very useful for the progress of computer science. Before Chomsky, a program to be executed by a computing machine, concrete or abstract, was assumed to be entirely a mathematical issue. It was so for Lovelace (1843), celebrated as the first programmer in history (Essinger, 2014), and it was also the case for Turing (1936) (see §2.2.1) and other computer pioneers during the first half of the past century (Hopper, 1981). In between the end of the fifties and the beginning of the sixties, a radical new perspective occurred in the practice of software programming, with Chomsky's linguistics taking over the prominent role previously held by mathematics. The majority of the historical accounts of programming have focused on reasons behind the popularity of the linguistic appeal: from the needs of machine independence and generality (Gobbo and Durnová, 2014), to specific features of the early programming languages. By these analyses, computing causally shifted from a mathematical to a linguistic paradigm stimulated by essential necessities. The need for code flexibility, accessibility, and speed led to the familiar programming languages. However, little attention seems to have been paid to the very notions that made such an outcome possible. In the early 1960s, several crucial elements of Chomskian linguistics became so widespread as to be conceived as the natural way to deal with formal languages and then, to the particular kind of formal languages that programming involves. This historical coincidence certainly contributed, in the words of Nofre et al. (2014), "to an increased understanding of programming as a machine–independent activity", and therefore "programming languages became new epistemic objects that were no longer connected to a particular machine [...]". But these same notions were, for the lack of an appropriate perspective, an impediment to understand the full implications of natural language for software, as discussed later in this paper. Programming languages were described in a clear analogy to Chomsky's rules. John Backus, who introduced FORTRAN, the first programming language, wrote:

> In the description of IAL syntax which follows we shall need some metalinguistic conventions for characterizing various strings of symbols. To begin, we shall need metalinguistic formulas.

(Backus, 1959, p.14)

Some examples of metalinguistic formulas for integers and numbers by Backus are:

$$\langle \text{digit} \rangle \quad :\equiv \quad 0\,\overline{\text{or}}\,1\,\overline{\text{or}}\,2\,\overline{\text{or}}\,3\,\overline{\text{or}}\,4\,\overline{\text{or}}\,5\,\overline{\text{or}}\,6\,\overline{\text{or}}\,7\,\overline{\text{or}}\,8\,\overline{\text{or}}\,9 \tag{3.8}$$

$$\langle \text{integer} \rangle \quad :\equiv \quad \langle \text{digit} \rangle\,\overline{\text{or}}\,\langle \text{integer} \rangle\,\langle \text{digit} \rangle \tag{3.9}$$

$$\langle \text{dn} \rangle \quad :\equiv \quad \langle \text{integer} \rangle\,.\overline{\text{or}}.\,\langle \text{integer} \rangle\,\overline{\text{or}}\,\langle \text{digit} \rangle\,\langle \text{dn} \rangle\,\overline{\text{or}}\,\langle \text{digit} \rangle \tag{3.10}$$

In these kind of formulas, $\overline{\text{or}}$ means possible alternative values; for example, a digit may have any decimal digit value from 0 to 9. Formulas may be recursive – say integer (in the left hand side) is expressed in terms of integer (in the right hand side). The symbol 'dn' stands for decimal number. Despite differences in notation, Backus' formulas are exactly rewriting rules, and use terminal and non-terminal elements. Chomsky always remained unimpressed and disinterested with respect to the use of his theoretical apparatus in computer science.

3.2.3 *Understanding Natural Language*

This section borrows the title of Terry Winograd's book (1972) that portrays the first stunning achievement of AI based on Chomsky. The fallout from Chomskyan linguistics in computer science, seen in the previous section, was indirectly important for AI too, for the trivial reason that AI relies on programming languages. But there was a stronger motivation for AI to look at the work of Chomsky: the mathematical treatment of language could have been the way to endow the computer with linguistic capability. Chomsky's attitude towards this effort, and to AI in general, was even worse than his attitude to computer science:

> [...] the mathematical theory of communication, cybernetics [...] were in a period of rapid development and much exuberance. Their contribution lent an aura of science and mathematics to the study of language and aroused much enthusiasm. [...] My personal reaction to this particular complex of beliefs, interests, and expectations was almost wholly negative. [...] The models of language that were being discussed and investigated had little plausibility [...] and I had no personal interest in the experimental studies and technological advances. The latter seemed to me in some respect harmful in their impact.
>
> (Chomsky, 1975, p.39)

In addition to intellectual reasons for distancing from AI, there appears to have been, in the 1960s, some unpleasant rivalry between Chomsky and the head of the MIT AI Lab, Marvin Minsky, both young, brilliant, and very ambitious (Boden, 2008, p.673).

It was Minsky himself who launched in the mid-1960s, a project at MIT in which Chomsky's theories would find their place. It was one of the first attempts in AI to use robots in testing ideas about the nature of intelligence. The robot, at the request of an interlocutor, would have to pick up blocks from a table, build them into towers, or put them into and out of a box (Minsky and Papert, 1972). Incidentally, in planning this project, Minsky made a memorable estimation error about the difficulty in artificial vision. He evaluated that the image processing component of the robot could be programmed in a summer by undergraduate students. A summer vision project was then established in 1966, with the goal of naming objects found in an image by matching them with a predefined vocabulary (Papert, 1966), a project doomed to failure.

In contrast to his thoughts on the vision component, Minsky was very shrewd for the linguistic segment, which he committed to Winograd as doctoral research. He could not have made a better choice. It was a huge success, and Winograd's thesis was immediately published in book form by a leading university press (Winograd, 1972). The robot communication program developed by Winograd was called *SHRDLU*, not an acronym but a shrewd joke: they were the first letters in the keyboard of a standard linotype typesetting machine of the 1960s. The sequence *SHRDLU*, in a book, was a mark used by typesetters so that the proof-readers would easily spot mistakes (Boden, 2008, p.685).

A fundamental component of SHRDLU was the syntactic analysis based on Chomsky's abstract grammar. Winograd highlighted this fundamental difference with respect to the hasty artificial conversation attempts like Eliza (see §2.3.2):

> Computer programs for natural language took two separate paths. The first was to ignore traditional syntax entirely, and to use some sort of more general pattern matching process to get information out of sentences. Systems such as [...] ELIZA (Weizenbaum, 1966) [...] made no attempt to do a complete syntactic analysis of the inputs. They either limited the user to a small set of fixed input forms or limited their understanding to those things they could get while ignoring syntax.
>
> The other approach was to take a simplified subset of English which could be handled by a well-understood form of grammar, such as one of the variations of context-free grammars. There has been much interesting research on the properties of abstract languages and the algorithms needed to parse them.
>
> (Winograd, 1972, p.42)

The alternative approach mentioned by Winograd was an active area of research in Chomskyan linguistics, aimed at finding algorithms for constructing syntactic trees of a given sentence. These algorithms are called *parsers*, and can follow two different strategies, as sketched in Figure 3.4. Parsing always starts from a tree where the only known nodes are the root S and the leaves, made by the terminal elements of the input sentence. The *bottom-up* strategy proceeds shifting pairs of leaves from left to right into the parsing mechanism, then checking if the two terminal elements can be reduced in a non-terminal category by a rewriting rule of the grammar. Due to this process, these parsers are also called *shift-reduce*. One of the earlier *bottom-up* algorithms is known as *CKY*,

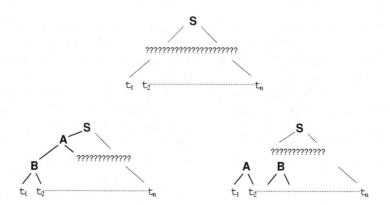

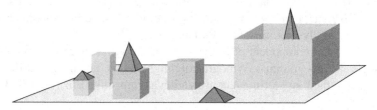

Figure 3.4: The two parsing strategies, with respect to the problem of parsing illustrated at the top: the *top-down* strategy in the left plot; the *bottom-up* strategy in the right plot.

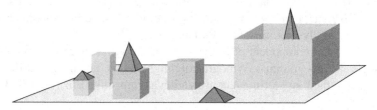

Figure 3.5: The micro-world SHRDLU is able to talk about, composed of blocks, pyramids, and one box.

by the names of the three authors (Kasami, 1965). In the *top-down* strategy, the unknown nodes are filled by predicting which non-terminal expansions, starting from *S* and descending along the leftmost nodes, will match the terminal elements. For this reason, the *bottom-up* parsers are also called *predict-match*. One of the earlier algorithms in this category is thanks to Jay Earley (1970). Winograd adopted a top-down left-to-right parser.

Syntactic analysis is at the core of Winograd's system and, giving heed to Chomsky, this was all that was to be expected about understanding a language. Winograd went further, including a fairly sophisticated semantic analysis, integrated with world knowledge and even problem solving. The world to know, of course, was made of only blocks, pyramids, and one box. The problems to solve were to move and pile these simple objects. This micro-world, shown in Figure 3.5, was just a virtual abstraction of the simple environment that the robot would have, of which SHRDLU was the language interface. Here are some samples of SHRDLU's language competence (Winograd, 1972, pp.9–10):

human	how many blocks are not in the box?
SHRDLU	four of them
human	can a pyramid be supported by a block?
SHRDLU	yes
human	can the table pick up blocks?
SHRDLU	no
human	can a pyramid support a pyramid?
SHRDLU	I don't know

Moreover, SHRDLU is not all talk, it will try to execute – virtually – a requested task:

human	pick up a big red block
SHRDLU	OK [SHRDLU answers "OK" when it carries out a command]
human	grasp the pyramid
SHRDLU	I don't understand which pyramid you mean [there are 3]
human	find a block which is taller than the one you are holding and put it into the box
SHRDLU	by "it", I assume you mean the block which is taller than the one I am holding, OK

Clearly, the world known by SHRDLU is outrageously rarefied and the number of actions it could carry out and questions it could answer, extremely limited. Still, SHRDLU was a major breakthrough, with a linguistic competence such as to suggest that it actually 'understood' what was being asked. According to Dennett, the merits of Winograd go way beyond Chomsky:

> [...] Winograd's real contribution in SHRDLU is not that he has produced an English speaker and understander that is psychologically realistic at many different levels of analysis [...] but that he has explored some of the deepest demands on any system that can take direction (in a natural language), plan, change the world and keep track of the changes wrought or contemplated, and in the course of this exploration he has clarified the problems and proposed ingenious and plausible partial solutions to them.
>
> (Dennett, 1978, p.117)

After his PhD; Winograd never produced anything even remotely like SHRDLU; he even moved away from AI, and by the 1980s, became one of its critics. This wayward intellectual trajectory will be described later, in §4.4.3. In the meantime, Chomsky had, despite himself, broken into AI, and syntactic analysis had moved from abstract formulas to solving algorithms. A new domain was born, named NLP (*Natural Language Processing*) (Grosz et al., 1986; Allen, 1995; Dale et al., 2000). NLP inherited the rigor inherent in Chomskyan linguistics, with the advantage of fairly straightforward transfer from the theoretical domain to computational programs. At the same time, however, NLP inherited also, the unavoidable lack of coverage in generative grammars of many phenomena found in ordinary language. In the rigor of NLP, there is no more room for play, in particular for playing in Turing's 'imitation game', and software competing in the Loebner Prize (see §2.3.2) is easily treated with scorn. When AI began to lose credibility (see §5.1.1), many researchers in NLP did not consider themselves part of AI. In fact, NLP includes simple applications in language processing that do not aim to simulate intelligence and do not even require it.

3.3 Levels of Cognition

In the successes achieved in AI by Newell and Simon, and by Winograd, many of great significance to cognitive science, there lurked a hidden shadow: perception. When Newell and Simon chose Russell's theorems as the arena for their first problem solvers, it was a fallback from the initial task of playing chess because it involved visual perception. Winograd's SHRDLU was supposed to be the linguistic interface of the robot that never saw the light of day, because the perceptual component was a total misfire. This lack was particularly embarrassing to AI for two reasons: seeing does not seem to be a particularly complex ability, certainly less than proving theorems; moreover, the visual system was, already in 1960s, the best known part of the brain (Kuffler, 1953; Hubel and Wiesel, 1959, 1962; Hartline, 1967). The difficulties of AI with perception, visual in particular, were certainly due to the intrinsic complexity of this capacity, largely overlooked by AI, as seen in §3.2.3, but were exacerbated by AI's disinterest in neuroscience. We will discuss in §4.1 the strong tension that arose in AI with respect to neuroscience-inspired research.

The difficulties in covering aspects of cognition for which perception is highly relevant, and in correlating artificial processes with known brain structures, were embarrassing indeed. That is, until the moment when David Marr pointed out an elegant way out.

3.3.1 Brave Computational Explanations of the Brain

Marr began his research career as a doctoral student in theoretical neuroscience under the supervision of Giles Brindley at Trinity College, Cambridge. The question he chose to address in his work was both a general and an ambitious one, a theoretical speculation on how the brain works, meticulously compared with the anatomical data available at the time. Marr's venturing into theoretical neuroscience, decades before becoming established, certainly resulted from his personal genius, and was favored by his mathematical background, and his being involved in the experimental and applicative neuroscientific activity in the Physiological Laboratory in Cambridge led by Brindley, a pioneer in neural prostheses (Brindley and Lewin, 1968). The theoretical exploration of the brain with mathematical tools was not on Brindley's agenda, but was certainly within his insatiable scientific curiosity. He occasionally attended the Ratio Club, a British cybernetic dining club whose members included scientists such as Ross Ashby and Alan Turing (Holland and Husbands, 2011), as a guest.   The outcome of Marr's doctoral research is a trilogy with separate theories about how the brain works for each of the three major brain structures: archicortex (Marr, 1971), cerebellum (Marr, 1969), and neocortex (Marr, 1970). Common to the three theories is the concept of *codon*, a term Marr borrowed from molecular genetics, to address elements encoding subsets of features able to fire a neuron.

The neocortex is clearly the part of the brain where higher cognition takes place in humans, including the highest level of perception, and it was the most demanding target of Marr's enterprise. We will give a quick overview of his work here. More details can be found in (Plebe, 2018). The theory is described in a long, and difficult to read, paper (Marr, 1970), which remained one of his least known works.

The work of Marr is organized in two distinct parts. The first part is a description of the general problem faced by the cortex in pure abstract mathematical terms. The second part attempts to derive correspondences between entities of the mathematical model and cells in the cortex. The general task of the cortex is formalized as the ability of a sentient agent to classify an event E_i of the world as Ω_j, where each class Ω_j is a type of event of relevance for the organism. He used as an example of a class, the set of events that represent various types of poodles. The classification is based on a set of detectable features $a_{i,k}$, probabilistically associated with the event instance E_i. Two events that represent different poodles E_1 and E_2 should have more features $a_{1,k}$ and $a_{2,k}$ in common, than the number of features shared by E_1, and an event E_3 that represents something different from a poodle.

The basic law behind this problem is the so-called Fundamental Hypothesis:

> Where instances of a particular collection of intrinsic properties (i.e., properties already diagnosed from sensory information) tend to be grouped such that if some are present, most are, then other useful properties are likely to exist, which generalize over such instances. Further, properties often are grouped in this way.

> (Marr, 1970, p.182–183)

From this basic law, several theorems such as the 'diagnosis theorem', that relates the conditional probability $P(\Omega_j|E_i)$ of the event instance E_i to be of type Ω_j with the functions $c_k(E_i)$, called 'evidence function', returning 1 if the feature $a_{i,k}$ is present, 0 otherwise are informally derived. The 'interpretation theorem' deals with cases in which only a subset of features $a_{i,k}$ become available during an event instance.

After a detailed formulation of this general theory of how mathematical models might explain classification of sensory signals, Marr engaged in an audacious attempt to adapt the working of the cortex to this theory.   He speculated that Martinotti (1890) cells may be plastic *codon* cells, and spiny stellate cells in layer IV could be codons with predefined output classes of events. He provided several diagrams of template circuits, based on Martinotti, spiny stellate, and pyramidal cells, that fulfilled his theory. Marr completed the mapping with a series of schemes of the interconnections between the neural elements, where their causal connections matched the mathematical relations in his model. For example, there were pyramidal cells whose output matched with the function $P(\Omega_j|c_i)$, with c_i the activation of an excitatory codon cell synapsing to the pyramidal cell. Given the complexity of the cortex, and the frugal amount of details available at the beginning of the 70s, Marr's attempt to derive an organization at the level of neural circuits just from abstract mathematical laws was hopeless. In the period between 1971 and 1972 Marr acknowledged the risks run by abstract general theories of the brain that lacked an understanding of specific neural mechanisms. The risk was that of results, glaringly incomplete and almost fruitless.

3.3.2 Goodbye Neuroscience, Welcome AI

His shift in view is clearly stated in a review of the proceedings of a summer school held in Trieste in 1973:

> Many experimental biologists dismiss with contempt the approach of even very able theoreticians in developmental or neurophysiological problems. The outsider need look no further than this volume to understand why. [...] papers describe attempts to elucidate problems of biological information processing, but in one way or another they all make the same strategic error – engaging in the search for a general theory before and actually instead of tackling the particular problems at hand. [...] With problems of biological information processing there has been almost no experience, and one's intuition is at best untrustworthy. It may even be that biological information processing admits of no general theories except those so unspecific as to have only descriptive and not predictive powers.
>
> (Marr, 1975, p.875)

In this period, Marr himself was following the direction recommended in this review. When he moved, in 1973, to the Artificial Intelligence Laboratory at MIT, he abandoned the speculative theoretical road, shifting his efforts to the study of the visual system, favoring bottom-up work grounded in an understanding of mechanisms involved in specific tasks. At MIT, and under the influence of Minsky, Marr developed an account of cognition based on levels, crucial in stipulating an agreement between cognitive and neuroscientific aspects. The distinction in levels was developed for visual perception, but was later generalized to psychology as a whole.

In the vision domain, the separation in levels of description was already pursued by Tomaso Poggio, a colleague of Marr at MIT. In a paper exploring the fly's visual system, Reichardt and Poggio (1976) established three levels of analysis:

1. a 'phenomenological' level, concerning the overall function of the system, its input-output behavior, and its logical organization;

2. a 'functional' level dealing with the functional principles of the subsystems;

3. a third level concerned with individual components and detailed circuitry of the system.

In a joint paper, Marr and Poggio (1979) identified the following four different levels, without specific relation to vision:

1. that at which the nature of a computation is expressed;

2. that at which the algorithms that implement a computation are characterized;

3. that at which an algorithm is committed to particular mechanisms;

4. that at which the mechanisms are realized in hardware.

Eventually, the levels adopted by Marr in his famous book, *Vision*, are back to three again, and respond to the following different questions (Marr, 1982, p.25):

1. computational level: What is the goal of the computation, why is it appropriate, and what is the logic of the strategy by which it can be carried out?

2. algorithmic level: How can this computational theory be implemented? In particular, what is the representation for the input and output, and what is the algorithm for the transformation?

3. hardware level: How can the representation and algorithm be realized physically?

Marr insisted in a relative independence of the three levels, and highlighted the possibility that some phenomena may be explained at only one level. He also gave the computational level priority, with questions at the other levels derived in light of analyses at the first. The independence of the computational level allowed Marr to develop a model of visual perception with flexible degrees of correspondence with neurophysiology. His theory of how, in earlier processing stages, edges are detected is formally derived from mathematical optimality requirements, but also taking into account the known computational properties of cortical areas where edge detection takes place (Marr and Hildreth, 1980). Instead, Marr's theory of representation of three dimensional objects is purely derived from the requirements of the problem and its constrains. The matter, widely debated, concerns how to reconcile the two-dimensional topological organization of the retina and the visual cortex with recognizing objects independently of their arrangement in three dimensions. Marr postulated that there is an intermediate representation he called '$2\frac{1}{2}$–D sketch', made on 2D surfaces with associated vectors normal to the surfaces, and a final symbolic representation of objects by primitive elements, called *generalized cylinders* (Marr and Nishihara, 1978; Marr, 1982). No trace was ever found in the visual system of either the '$2\frac{1}{2}$–D sketch' or the generalized cylinders, but this was not a particular worry, given the independence of the computational level.

The theoretical priority of the computational level offered cognitive science an elegant way out from the intricacies of neurophysiology. This opportunity was soon championed, among others, by the Canadian psychologist Zenon Pylyshyn (1981, 1984), heralding the autonomy of the computational level, and by Fodor (1974), advocating the full autonomy of psychology from neuroscience.

Chapter 4

Contending Philosophical Frameworks Within Artificial Intelligence

A remarkable feature of AI is the plurality of different philosophical views. We can find in the AI community, the coexistence of different wisdoms encompassing assumptions about human intelligence and about the best ways to achieve artificial intelligence. Most of those wisdoms have a direct correspondence with well established philosophical traditions. The first steps of AI and its early influence on cognitive science, narrated in the previous chapters, were dominated by the rationalist tradition. Soon, however, other philosophical guises found sympathy in the AI community, establishing a variety of different views. Sometimes the rivalry between contending philosophical views escalated to distinct sociological coteries, in harsh competition for intellectual and economic resources. This happened between the well established rationalist part in AI and the other most important component, empiricism. The rivalry between rationalism and empiricism in AI is not too surprising, it just mirrors a much longer conflict between the same two parts in philosophy, and it has fueled a similar contention in cognitive science. Alongside rationalism and empiricism, other philosophical guises can be found in AI including Darwinism, Heidegger influences, and even flavours of Eastern philosophical traditions.

4.1 The Empiricist Agenda

The empiricist view in artificial intelligence connects to a long standing philosophical tradition that originated with Aristotle, who insisted that knowledge is based on experience rather than innate ideas. The empiricist programme was revived in the seventeenth century by some British philosophers, first by John Locke (1690), claiming that mind comes into the world like an "empty cabinet", then by David Hume (1739), who argued that concepts are the product of what we perceive, and reason gets all its materials from perception. The battle between rationalism and empiricism has endured through the long history of Western philosophy because of radically different views. For rationalists, knowledge is built up on a scaffolding of innate principles while for empiricists, all knowledge derives from experience.

For most, if not all, scholars of AI, Alan Turing (1950) with his celebrated *Computing Machinery and Intelligence* set the stage of the discipline (see §2.2). Not many, however, know that Turing, two years earlier, had also initiated the empiricist agenda for AI with the report, *Intelligent Machinery* (1948b). He envisioned a machine based on distributed interconnected elements, called *B-type unorganized machine*. Turing's neurons were simple NAND gates with two inputs, randomly interconnected; each NAND input could be connected or disconnected, and thus a learning method could 'organize' the machine by modifying the connections. His idea of learning generic algorithms by reinforcing useful links and by cutting useless ones was the most farsighted of this report, anticipating the empiricist approach in AI. Turing made his commitment to empiricism concerning the human mind explicit:

> We believe then that there are large parts of the brain, chiefly in the cortex, whose function is largely indeterminate. [...] All of this suggests that the cortex of the infant is an unorganised machine, which can be organised by suitable interfering training.
>
> (Turing, 1948b, p.16)

This report was commissioned by the UK National Physical Laboratory, which board was not as farsighted as Turing; the work was dismissed as a 'schoolboy essay'. Therefore, this report remained hidden for decades, until upheld by Copeland and Proudfoot (1996).

4.1.1 The Perceptron and its Backlash

Orphaned of Turing's contribution, the early empiricist attempts in AI did not use digital computers. Rather, they followed the road paved by the blending of electrical and electronic engineering with neurophysiology in the first half of last century; this is recognized as significant in shaping neuroscience research (Brazier, 1961; Newcomb, 1994; Rose and Abi-Rached, 2013). The Schmitt trigger is a basic circuit well known to every student in electronic engineering, Otto Schmitt built it in 1936 for the purpose of simulating the behavior of a neural membrane. Other neural inspired electronic devices were the Neuromime (Harmon, 1959), and the Neuristor (Crane, 1960). This wave of enthusiasm galvanized even the young Marvin Minsky (1954), who engaged in the construction of the first artificial neural network for his PhD project. The machine, named SNARC (*Stochastic Neural Analog Reinforcement Computer*), assembled 40 artificial neurons, each made with six vacuum tubes and a motor to mechanically adjust its connections. The objective of SNARC was to try and find the exit from a maze where the machine would play the part of a rat. Running a mouse through a maze to investigate behavioral patterns was one of the most common paradigms in empirical psychology at that time (Tolman, 1948). The construction of SNARC was an ambitious project, the first design of a learning machine in an empiricist vein, and one of the first machines aiming at imitating the brain at neural level. However, it had little influence on AI, and soon Minsky turned to rationalism, partly due to the influence of John McCarthy, with whom he founded the AI Lab at MIT, a flagship of rationalism.

The most influential project at that time in the empiricist agenda was the 'perceptron', an electronic device designed by Frank Rosenblatt (1958, 1959, 1962) at the Cornell Aeronautical Laboratory. The general scheme of the perceptron is given on the left side of Figure 4.1. The first prototype, Mark I Perceptron, had the S-point layer made of 400 photosensitive units arranged in a 20×20 retina matrix, the A-units layer of 512 units, and the R-units layer made of 8 response units. Each A-unit receives excitatory connections from several S-units,

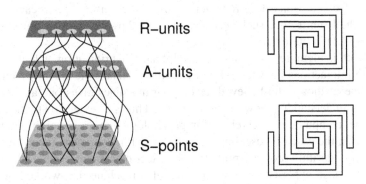

Figure 4.1: On the left, the scheme of Rosenblatt's perceptron. The 'S-point' assembly is a retina-like matrix of sensory elements, 'A-units' is a vector of associative units taking connections from S-point, and 'R-units' is the vector of response units, with one R-unit for each class of object to be recognized in the retina. On the right, the two drawings on the cover of the book by Minsky and Papert. While looking similar, the two spirals are topologically different. The top one is two separate but intertwined curves, the bottom one is a single closed curve. Minsky and Papert demonstrated that no perceptron machine can distinguish the two drawings.

and the unit fires if the sum of the signals from input connections reaches a fixed threshold. The layer R is coupled to layer A in a similar way, with an important difference: connections from A-units to R-units can change dynamically, using motor-driven potentiometers. That is where the learning is hidden.

There are a number of key concepts in the enterprise of Rosenblatt which, together, characterize future aspects of AI. Learning in the perceptron is realized as close as possible to how it works in the brain. Among the various modes of learning in the brain, only the one that was best known at the time was adopted: synaptic plasticity. This, today, is still the main choice in the models that follow the empiricist agenda in AI (see §5.1.2). The interpretation of synaptic plasticity in the perceptron was necessarily a crude simplification, implemented by the following rule:

$$w_{i,j}^{(t+1)} = w_{i,j}^{(t)} + \eta a_i \left(\hat{r}_j^{(t)} - r_j \right) \tag{4.1}$$

where a_i is a unit in layer A connected with an R-unit r_j by a synapse with efficiency represented by the real number $w_{i,j}$, with a typical range $[0..1]$. When the t-th sample is presented to the S layer, the correct response pattern is known, and $\hat{r}_j^{(t)}$ is the correct level of activity of the unit r_j for the t-th sample. By virtue of rule (4.1), if a unit in the R layer fires although it should not, the connection weights of all currently active A-units to this R-unit are reduced. Conversely, if the unit does not fire although it should, the connection weights of all currently active A-units to this R-unit are increased. In the case that the unit a_i in layer A is inactive for the t-th sample, no modification will occur. In his later theoretical book, Rosenblatt (1962) was able to demonstrate mathematically, the power of learning in the perceptron. His convergence theorem states that if there exists a vector $\vec{w}$ of connections $A \to R$ which leads to the correct classification of patterns by the R-units, then such vector can be found by the learning procedure of Equation (4.1) in finite time.

Rosenblatt inaugurated an unusual disciplinary hybridization, which will mark future interactions between AI and cognitive science. He was a psychologist, but one of those who believed developing machines gives one an added dimension of understanding how the mind works. This is an extension to the discovery of the computer as research tool in psychology that was emerging in the early 1960s. In the case of Rosenblatt, psychological investigation involved not only abstract algorithm programming, but also electronic engineering design.

Rosenblatt also experienced how the idea of a machine that worked quite like the brain received a lot of press attention. When, on 7 July 1958, the US Office of Naval Research, founder of the perceptron project, presented it at a press conference, *The New York Times* reported the event using these words: "The Navy revealed the embryo of an electronic computer today that it expects will be able to walk, talk, see, write, reproduce itself and be conscious of its existence." The media excitement for brain-like devices has endured all along the history of AI, up to the current days.

Being a psychologist, and being popular in media coverage, made Rosenblatt the target of Minsky's wrath. According to Robert Hemt-Nielsen:

> During this era of the early 1960s, a significant group of researcher with a less "biological" and more computer science-oriented approach [...] found it emotionally intolerable that some of what they saw as "lightweight" research on neural networks was receiving a lot of press attention . A particularly dramatic case of this was Marvin Minsky. Minsky had gone to the same New York "science" high school as Frank Rosenblatt, a Cornell psychology Ph.D. whose perceptron neural network pattern recognition machine was receiving significant media attention. The wall-to-wall media coverage of Rosenblatt and his machine irked Minsky. One reason was that although Rosenblatt's training was in "soft science", his perceptron work was quite mathematical and quite sound – turf that Minsky, with his "hard science" Princeton mathematics Ph.D. didn't feel Rosenblatt belonged on.
>
> (Anderson and Rosenfeld, 2000, p.303)

Ironically, Minsky started out as one of the earliest pioneers in neural hardware, and for his SNARC machine he introduced technical solutions, like the motor-driven potentiometer, adopted in the Mark I perceptron too. But he soon turned to the rationalist side of AI and become the strongest adversary of the empiricist side. Minsky involved a leading researcher at MIT, Seymour Papert, in the project of a highly elaborated mathematical analysis of the perceptron, with the only purpose of discovering and highlighting all its limitations. In fact, Rosenblatt was already aware of several limitations of the perceptron, discussed at length in his theoretical book (Rosenblatt, 1962), and the work of Minsky and Papert just added more examples to the already known limitations.

For a few years in the 1960s, Minsky and Papert presented their ever-growing body of mathematical results at conferences to discredit empiricist neural research. Later, their effort culminated in a published book, with title *Perceptron*, where they explained their intent in clear terms:

> Both of the present authors (first independently and later together) became involved with a somewhat therapeutic compulsion: to dispel what we feared to be the first shadows of a "holistic" or "Gestalt" misconception that would threaten to haunt the fields of engineering and artificial intelligence as it had earlier haunted biology and psychology.
>
> (Minsky and Papert, 1969, p.19)

The cover of the book illustrated an allegedly striking case of failure for the perceptron, here shown in Figure 4.1 on the right. Even this limitation due to difference in topology, had, in fact, already been pointed out by (Rosenblatt, 1962, p.579). Replication of results coming from an area one is not directly involved with, with the only purpose of discrediting another's research, are unusual in science, and the book of Minsky and Papert become a case in sociology of science

(Olazaran, 1996). The main reason for the attack on Rosenblatt was the competition for research funding; the rationalist community was worried by a diversion of resources to the new neural research area. For this purpose it was crucial to disparage the entire domain of neural network research, not just the perceptron, as done in several passages of the book, as in these discouraging words:

> The perceptron has shown itself worthy of study despite (and even because of!) its severe limitations. It has many features to attract attention: its linearity; its intriguing learning theorem; its clear paradigmatic simplicity as a kind of parallel computation. There is no reason to suppose that any of these virtues carry over to the many-layered version. Nevertheless, we consider it to be an important research problem to elucidate (or reject) our intuitive judgment that the extension is sterile.
>
> (Minsky and Papert, 1969, p.231)

The attack was fully successful, the book convinced a large part of the AI community. In the years which followed, there were books on data representation, like (Simon, 1986), in which the section about perceptron is titled *The birth and death of a myth*. Above all, the attack brought the desired results: founding for the perceptron project dried up, Minsky was awarded of more than ten million dollars for the development of rationalist AI at MIT.

Shortly afterwards, on his forty-third birthday on July 11 1971, Rosenblatt died in a boating accident, and there were rumors that he might have committed suicide.

4.1.2 The Magic of Backpropagation

Rosenblatt had not merely identified the limitations of the perceptron, he had also diagnosed the source of the criticality, indicated the path to overcome them. What he called 'simple perceptron', like the Mark I machine, had its inherent limitations due to a single layer of adaptable connections, from the A-units to the R-units. The solution he envisioned was simply to allow connections from the S-units to the A-units to adapt as well. The difficulty is that the simple rule of Equation (4.1) cannot be extended to the $S \rightarrow A$ connections, because there is no external 'teacher' that tells which unit in the A layer should fire or not for an input pattern. Rosenblatt sketched the following solution:

> The procedure to be described here is called the "back-propagating error correction procedure" since it takes its cue from the error of the R-units, propagating corrections back towards the sensory end of the network if it fails to make a satisfactory correction quickly at the response end.
>
> (Rosenblatt, 1962, p.292)

The concept and the term 'backpropagation' introduced by Rosenblatt in this passage were doomed to be the holy grail of the empiricist agenda in AI right up to the present day. The missing piece was an algorithm realizing the back-propagating error correction procedure presented by Rosenblatt in descriptive terms. It was found twenty years later by Rumelhart et al. (1986).

The artificial neural model for which backpropagation applies, called the *feedforward network*, is made of simple units organized into distinct layers, with unidirectional connections between each layer and the next one. The values of the units are computed with the following equations:

$$\vec{x}_1 = \mathbf{A}^{(I)}\vec{x} + \vec{b}^{(I)}, \tag{4.2}$$

$$\hat{f}(\vec{x}) = \mathbf{A}^{(O)}\vec{x}_N + \vec{b}^{(O)}, \tag{4.3}$$

$$x_{i,k} = h\left(\vec{w}_{i,k}\vec{x}_{i-1} - \theta_{i,k}\right) \quad 1 < i < N. \tag{4.4}$$

The first layer, ruled by Equation (4.2), simply provides input values to the network, normalized with the linear operators $\mathbf{A}^{(I)}$ and $\vec{b}^{(I)}$. The top layer is where the output data appear. The entire feedforward network can be expressed as a function $\hat{f}(\vec{x})$ of the input vector $\vec{x}$ and, as described by Equation (4.3), it is normalized again to meet the desired data range. The real dirty work is done by the layers in between, according to Equation (4.4). In a layer i, each unit $x_{i,k}$ sums up all the values from the previous level, weighted by parameters $\vec{w}_{i,k}$, minus a threshold $\theta_{i,k}$, and the result is modified by a non-linear activation function $h(\cdot)$, such as the sigmoid or the hyperbolic tangent. A sketch illustrating the computation in feedforward networks is provided in Figure 4.2.

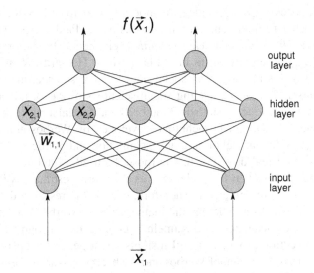

Figure 4.2: Sketch of an artificial neural feedforward network. The mathematical symbols correspond to those used in Equations (4.2), (4.3), (4.4) in the text.

Comparing the feedforward network in Figure 4.2 with the perceptron in Figure 4.1, the similarities are striking. The first layer in the feedforward network is now a general vector, without the retina-like two dimensional arrangement of units of the perceptron's S layer. The layer in between the input and the output layers, formerly called A layer, is now the 'hidden' layer. This name is thanks to Geoffrey Hinton, one of the father of feedforward networks, when he first came to know and study hidden Markov models, and decided 'hidden' was good for the intermediate layer in the network (Anderson and Rosenfeld, 2000, p.379). The term is appropriate because, for the input and output layers, it is straightforward to assign a meaning to each unit, dictated by the nature of the problem to be solved by the neural network. Conversely, there is no overt meaning that can be assigned to the individual units in the 'hidden' layer.

The important difference between the new feedforward networks and the perceptron is that all connections are plastic; they adapt following the brand new backpropagation learning algorithm. The vector of all learnable parameters in the network being $\vec{w}$, and $\mathscr{L}(\vec{x}, \vec{w})$ a measure of the error of the network with parameters $\vec{w}$ when applied to the sample $\vec{x}$, the backpropagation updates the parameters iteratively, according to the following formula:

$$\vec{w}_{t+1} = \vec{w}_t - \eta \nabla_w \mathscr{L}(\vec{x}_t, \vec{w}_t) \qquad (4.5)$$

where t spans all available samples $\vec{x}_t$, and η is the *learning rate*. Note that in order to compute the error $\mathscr{L}((\vec{x}, \vec{w}))$, the correct responses to all samples $\vec{x}$ used for learning should be known; this learning strategy is generally called *supervised learning*.

In fact, the backpropagation idea was not only anticipated by Rosenblatt, it also had an earlier formulation, but without this name. Paul Werbos (1994), by titling his book *The Roots of Backpropagation*, claimed to be the originator of the backpropagation algorithm, in his PhD thesis at Harvard (Werbos, 1974).

The domain of his research was social and political science, his supervisor was Karl Deutsch, one of the leading social scientists of the 20th century, and one of the first in introducing statistical methods and formal analysis in political and social sciences. The novel technique developed by Werbos aimed at testing the Deutsch-Solow model of national assimilation and political mobilization (Deutsch, 1966) on real data. For this purpose, he used an iterative technique, termed *dynamic feedback* in which derivatives of the error estimates with respect to the parameters were computed. Therefore, even if the research domain was far away from networks and brains, the mathematics of Werbos was much convergent with backpropagation, and Rumelhart et al. (1995) recognized that this independent invention provided useful insights about general properties of the algorithm. Both the algorithm of Werbos and the backpropagation formulated by Rumelhart and Hinton have their roots in the gradient methods, developed in mathematics for the minimization of a continuously differentiable function $f(x)$. The first proposal of a gradient method dates back to Augustin-Louis Cauchy

(1847), and found renewed interest in the first half of the last century in several engineering applications. The Fire Control Design Division, the advanced military research division at the Frankford Arsenal, was where high technologies such as LIDAR have been invented. Two notable mathematicians in this departments, Levenberg (1944) and Curry (1944), developed two independent refinements of the original method of Cauchy; further variants, including his own, are collected in Polak (1971). Therefore, there was a mature mathematical context in the '70s, for using gradient methods in solving engineering problems, fertile enough to be imported in the domain of artificial neural networks.

Backpropagation had the magic to dissolve in one shot, the heavy limitations that Minsky and Papert had prophesied would plague all neural networks. Quite soon, a series of mathematical studies (Cybenko, 1989; Hornik et al., 1989; Stinchcombe and White, 1989) provided new demonstrations of the learning power of feedforward neural networks with at least one hidden layer. We saw earlier that the convergence theorem of the perceptron required the existence of a vector $\vec{w}$ of connections $A \to R$ which leads to the correct output function. In general, this vector may simply not exist. The theorems for feedforward neural networks ensure that a set of weights does exist for a large family of functions that map input to outputs, and this set can be approximated by learning. In other words, feedforward neural networks empowered by backpropagation are a universal approximator.

4.1.3 Commitment to Cognition, Not so Much to Neuroscience

The invention of backpropagation was the tip of the iceberg of an extensive research effort in artificial neural networks by a community identified as PDP Research Group, *Parallel Distributed Processing*. Their manifesto was a two-volume collection of articles edited by Rumelhart and McClelland (1986b). The title of this book, *Parallel Distributed Processing: Explorations in the Microstructure of Cognition*, leaves no doubt about the commitment of this research group to the study of cognition. The new models made by artificial neurons that learn by backpropagation are proposed, first and foremost, as vehicles for understanding cognition. These models will be referred to here as ANN (*Artificial Neural Networks*), as they are usually named, together with other terms as PDP models or *connectionist* models. For many leading figures of the PDP Research Group, the strong interaction with cognitive science was a natural consequence of their disciplinary background. As in the case of Rosenblatt, scholars such as Geoffrey Hinton, Michael Jordan, James McClelland, David Rumelhart, and Timothy Rogers were all psychologists.

The final goal of Rosenblatt was a theory of pattern perception, supported by a working machine that imitated biological neurons. A very ambitious project, but the 'explorations in the microstructure of cognition' aimed to extend far beyond perception. The PDP project aimed at transforming the landscape of cognitive science under an empiricist view. The standard account of cognition before the PDP turn assumed that mature cognition develops from initial, innately specified and domain-specific knowledge, structured as discrete rules. Development involves relatively discrete, sudden transitions from one pattern of performance to another, when a new rule is discovered. ANN models can simulate aspects of cognitive behavior that do not implement explicit rules, and emerge from domain-general experience. Backpropagation allows neural models to link patterns of developmental change to specific aspects of the structure of the environment, and to show how experience with that structure will alter processing in the system. The primary arena of confrontation between the classical rule-based rationalist cognitive science and the PDP empiricist alternative is language development and that deserves its own section, 4.2.

While the PDP Group forged an even stronger alliance with cognitive science than Rosenblatt, the group lost connection with neuroscience. Indeed, most protagonists of the PDP Group were fond admirers of neuroscience, and they often have described their approach as inspired from neuroscience as seen, for example, in (Rogers and McClelland, 2014, p.1032–1034). However, it was unrequited love. James Bower (Miller and Bower, 2013, p.5) recalls one of the first neural network meetings at Santa Barbara in 1983, where participants "represented a remarkable mix of scientists and government officials [...] only two made any claim to being real biologists, myself and Terry Sejnowski. [...] I presented my work with Matt Wilson modeling the olfactory cortex and I remember distinctly that it was news to many in the room that synaptic inputs could also be inhibitory." During the PDP project, attempts were made to involve neuroscientists in the computational world; a good example was the series of conferences NIPS (Neural Information Processing) started in 1986. For James Bower (Miller and Bower, 2013, p.6) "the neurobiologists, including my friend John Miller, who I had invited to participate in the second NIPS meeting, found most of the talks either irrelevant to neurobiology or naive in their neurobiological claims." On the other hand, in neuroscience, there was a genuine interest in exploring the nature of the processing tasks executed by nerve cells and systems with computations, that could not be fulfilled by the simple PDP models. For this aim, a new field was established, called *Computational Neuroscience* or sometimes *Theoretical Neuroscience* (Dayan and Abbott, 2001), with its own series of conferences, like those which CNS started in 1992 (Miller and Bower, 2013).

A major breakthrough in computational neuroscience was the development of two distinct neural simulators: NEURON by Hines and Carnevale (1997) and GENESIS by Bower and Beeman (1998). Both computational frameworks provide environments for implementing in software, biologically realistic models of electrical and chemical signaling between neurons. The emergence of the PDP project was between 1977 and 1998, but the two researches had no interaction, with PDP-style networks and computational neuroscience taking two diverging paths. Another influential line of research within computational neuroscience was aimed at designing the so-called *canonical microcircuits* of the cerebral cortex (Shepherd, 1988; Douglas et al., 1989; Douglas and Martin, 2004; Plebe, 2018). This research focused on the highly repetitive circuital structure of the cortex and attempted to identify a sort of prototype circuit and corresponding governing equations, able to explaining its peculiar computational efficiency. There has been almost no cross-fertilization between computational neuroscience and the community of PDP-style artificial neural networks.

In a sense, the detachment of the ANN research from neuroscience has been beneficial for its acceptance in cognitive science. In the fierce debate on the reduction of psychology to neuroscience, defenders of functionalism argue for the autonomy of the former (see §3.3.2). On the opposite side, authors endorsing the reduction of psychological properties to brain properties reject functionalism altogether (Kim, 1992; Bickle, 2003). The PDP project accommodatingly eludes the controversy; ANN models are abstract enough for citizenship in the top level of Marr's hierarchy. However, their distributed structure is crucial in constraining the class of algorithms specified to perform a cognitive task (Churchland and Sejnowski, 1994). Therefore, the proposal of the PDP project is not alternative to the computational theory of the mind, it is still a form of computation, whose main difference from others derives precisely from the primacy of learning. In a traditional algorithm, the processing steps prescribe the function to be performed between the input and the output. In a backpropagation model there is no predefined function, the processing steps define how a general model will gradually adapt to any possible function by learning.

Some representatives of the PDP group would have appreciated stronger ties with neuroscience. One of them is Sejnowski, who fostered, often in alliance with the philosopher Patricia Churchland, the study of genuine brain computations for understanding cognitive behavior (Sejnowski et al., 1988; Churchland and Sejnowski, 1990, 1994).

In addition, Hinton, despite his fundamental contribution in the invention of backpropagation, did not like backpropagation that much. At the beginning he was even skeptical about its working: "I made one of these crazy inferences that people make – which was, that backpropagation is not very interesting [. . .] I had not really accepted at that point that this was the best you ever were going to be able to do," (Anderson and Rosenfeld, 2000, p.376). What Hinton dislikes in backpropagation is supervised learning, implying a sort of 'teacher' that provides the correct responses; this, of course, does not exist in the brain. His most loved model was an alternative to feedforward networks that used neither backpropagation, nor any sort of supervision, relying on the much more plausible *unsupervised learning* strategy. Hinton invented this model in collaboration with Sejnowski (Hinton and Sejnowski, 1983; Hinton et al., 1984; Hinton and Sejnowski, 1986), and called it the *Boltzmann Machine* (BM).

A scheme of the BM is shown in Figure 4.3, on the left. The units s_i are binary and stochastic, therefore can assume value 1 with some probability p_i. The

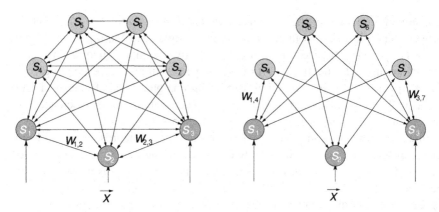

Figure 4.3: On the left, the scheme of a Boltzmann Machine, with the visible unit in dark and the hidden units in light gray. On the right, the same network as Restricted Boltzmann Machine, without interconnections between hidden units or between visible units.

BM conforms to the 'parallel distributed' paradigm with units connected by real value weights, but unlike feedforward networks, connections are reciprocal and symmetric. Some units are connected with external data and are called *visible units*, there are additional units, called *hidden units*, that capture the features 'hidden' inside the distribution of external data. The reason for the dedication to the great Austrian physicist is that a sort of 'energy' is defined for each unit, together with a dependency of the probabilities to a analogue to 'temperature'. For a unit s_i the gap of energy ΔE_i when the unit is on or off, is given by:

$$\Delta E_i = \vec{w}_i \vec{s}_i^{\mathrm{T}} \tag{4.6}$$

where $\vec{w}_i$ is the real vector of all weights connected to s_i, and $\vec{s}_i$ is the binary vector of all states of the units s connected to unit s_i. By using the gap of energy ΔE_i the unit s_i will be set to 1 with the following probability:

$$p_i = \frac{1}{1 + e^{-\frac{\Delta E_i}{T}}} \tag{4.7}$$

where T is the analogue of temperature. A BM reaches what is called 'thermal equilibrium', insisting on the metaphor with thermodynamics, when the distribution of the probabilities of all units will remain constant. Thermal equilibrium is reached easily by starting the evolution with high values of the temperature T in Equation (4.7), then gradually decreasing the temperature. A BM can reach thermal equilibrium either when an external binary vector $\vec{x}$ is applied to the visible units, or without external connections and letting the network run freely. These two possibilities are the key for a very simple learning rule. All connections $w_{i,j}$ between units s_i and s_j are changed after the presentation of a sample $\vec{x}$ by the following rule:

$$\delta w_{i,j} = \eta \left(p_{i,j} - \tilde{p}i, j \right) \tag{4.8}$$

where $p_{i,j}$ is the probability, at thermal equilibrium, for units s_i and s_j both being in the state on with $\vec{x}$ applied to the visible units, and $\bar{p}i,j$ is the corresponding probability when the network is running freely.

A simpler variant of the BM is the *Restricted Boltzmann Machine* (RBM), where there are no interconnections between visible units, and between hidden units; an example is shown in Figure 4.3 on the right. Both, the BM and the RBM, miss the output layer of feedforward networks, therefore it is less easy to retrieve the desired outcome of a trained model. One advantage of biological plausibility is that in the brain, it is not possible to identify neurons in a network as the only bearer of some 'output'. However, in a functionalist approach to cognition, it is very useful to identify precise inputs and outputs in a simulation model; this is one reason why the BM and the RBM have not had the same appreciation as feedforward ANN in cognitive science. Still, Hinton, talking twenty years later about the Boltzmann Machines, states, "I still think that's the nicest piece of theory I'll ever do," (Anderson and Rosenfeld, 2000, p.374), and we will see in §5.1.2 how Hinton was able to exploit the Boltzmann Machines again to invent the new wave of ANN.

4.2 The Case of Language Development

No area in cognitive science has been more shaken by the new wave of empiricism in AI than language development. As described in §3.2, Chomsky's linguistics has been of enormous importance to the history of cognitive science as a whole, but clearly, primarily, to the study of language. Yet a significant contradiction must be noted here. Although the Chomsky-inspired advances in the study of language evolution are, in a broad sense, evidence of the new wave of empiricism in AI, his theoretical intent was quite different. Animated by ambition and a spirit of contradiction to the typical theoretical background of the United States in the 1950s and 1960s, Chomsky aimed to establish a rationalist linguistics that stood in open opposition to the mainstream American philosophy of strict adherence to empiricism. For this reason, in 1966, he wrote, Cartesian Linguistics (Chomsky, 1966), in which he openly accepted the idea that language evolves on the basis of some innate structures, citing as his champion Wilhelm von Humboldt who had, indeed, freed himself from the opposition between empiricism and rationalism in the early nineteenth century and cultivated romanticism and linguistic idealism. Nevertheless, one of Chomsky's fundamental assumptions was that language is acquired thanks to a specialized computational apparatus with which the newborn's brain is already equipped. Various members of the PDP group have shown that it might be otherwise, that distributed systems of neuron-like units can learn language simply from experience. It was a provocation that triggered a broad, sometimes very acrimonious, debate that led to a profound impact of the empiricist AI side on cognitive science.

4.2.1 The Past Tense War

In many languages, inflectional morphology is the basic system for adapting words to express different grammatical categories such as case, person, number, gender, and tense. Inflectional morphological systems are often complex, therefore their acquisition is one of the most crucial challenges facing children. For this reason, this is one of the main subjects of investigation in developmental linguistics. In this context the acquisition of the past tense of English verbs is not, in fact, so special, but it has dominated the literature on inflectional morphology for thirty years.

The reason for this is because Rumelhart and McClelland (1986a) had targeted the English past tense formation in the first pure empiricist model of language learning, unleashing an epic war with the rationalists. The challenge that Rumelhart and McClelland launched was to show how learning the past tense morphology did not require any explicit and innate rules. Their work is chapter 18 of the PDP book (Rumelhart and McClelland, 1986b), and it is by far the most cited chapter. The model of Rumelhart and McClelland is shown in Figure 4.4: the 'neural' part is made of just two layers, the input layer with the phonological representation of the uninflected form of an English verb, and the output layer with the phonological form of the past tense of the same verb. This structure is more similar to the perceptron (see §4.1.1) than the new PDP style feedforward

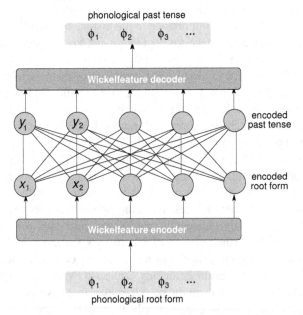

Figure 4.4: Scheme of the neural model of Rumelhart and McClelland (1986a) that learns the past tense of English verbs.

networks; there is no hidden layer, and the values of all units are binary: 0 or 1.

The activation of output units is, instead, borrowed from the BM (see §4.1.3), with the following probabilistic expression:

$$p_i = \frac{1}{1 + e^{-\frac{(\bar{x}_i \bar{w}_i i - \theta_i)}{T}}} \tag{4.9}$$

where p_i is the probability of the output unit i being on, $\bar{x}_i$ is the vector of all input units connected to the output unit i, and $\vec{w}_i$ is the corresponding vector of synaptic connection weights. As in the BM Equation (4.7), T is the analogue of temperature; with higher values, the response of the unit is highly variable. The way to learn in the model is exactly the same as in the perceptron:

$$w_{i,j}^{(t+1)} = w_{i,j}^{(t)} + \eta x_i \left(\hat{y}_j^{(t)} - y_j \right) \tag{4.10}$$

where $w_{i,j}$ is the connection between input unit i and output unit j, x_i is the value of the input unit i and y_j the value of the the the output unit j, while $\hat{y}_j$ is the value the output unit should have, to be the correct phonological representation of the past tense.

It may seem bizarre that this model, although part of the PDP project, does not use the crucial innovations introduced by the PDP with respect to the perceptron, like hidden layers and backpropagation. The reason is simple, Rumelhart and McClelland started to work on this model in 1982, when backpropagation was not yet invented. In fact, Rumelhart thought it would be necessary to use multiple layers, but there was a more serious problem to be solved first.

The major difficulty in using artificial neural models for language lies in the apparently irreconcilable diversity of the two formats. Language is an ordered sequence of auditory signals (in the spoken case) or symbols (in the written case), while a neural layer is a real vector with fixed dimension. Even for models limited to the processing of a single word, encoding an arbitrary length datum (the word), with a vector of fixed dimensions (the neural layer), is highly problematic.

Rumelhart and McClelland figured out a clever solution they called *Wickelfeature*, giving credit to Wickelgren (1969) who devised the scheme on which the solution is based. Wickelgren suggested that words can be represented as sequences of context-sensitive phoneme units, which represent each phone in a word as a triple, consisting of the phone itself, its predecessor, and its successor. This way, the coding has a fixed size for every possible word, with binary values that are on if a specific triple is contained in the word. The original scheme of Wickelgren lead to quite a long vector of dimension 42875. The *Wickelfeature* introduced by Rumelhart and McClelland is a reduction of all phonetic combinations into features capturing aspects of the central phoneme as well as its predecessor and its successor, ending up with a vector of 460 units.

There is a typical U-shaped profile of development observed in the acquisition of the English past tense (Kuczaj, 1977). Early in learning, children use highly frequent verbs which are mostly irregular, and they are inflected with their correct form. In a later phase of learning, children use a much larger number of verbs, the majority of which are regular. Now they make overregularization errors such as *comed* and *goed*, for the same irregular verbs that had previously been inflected correctly. At the final stage, the regular and irregular forms coexist. Rumelhart and McClelland trained their model with a corpus of verbs that simulates a young child's experience with past tenses, picking up English from everyday conversation (Francis and Kucera, 1967). The resulting errors during the learning run replicate with surprising similarity, the U-shaped pattern of development observed in children. Thus, the model had not only demonstrated the possibility of learning inflectional morphology without any innate rules, but also of doing so in a way similar to the error patterns observed in children.

This provocative result triggered a strong reaction from those cognitive scientists who adhered to Chomsky's rationalist view of language. Steven Pinker, one of the champions of the rationalist side, destined for a future fame no less than that of Chomsky, took charge of the defense. In a 102-page article, Pinker and Prince (1988) articulated a detailed critique of everything that could be called into question in the model of Rumelhart and McClelland. One of their most compelling arguments concerned the phonological representation used in the model, the *Wickelfeature*, a simplification that at the same time is scarcely plausible, and misses important aspects of the temporal order in phonological forms. Other less compelling criticisms were raised on the performances of the model and its rate of incorrect predictions.

For the rationalist component in cognitive science and linguistics, the critique by Pinker and Prince was fully successful, galvanized by the aggressive conclusive words of the article:

> We will conclude that the claim that parallel distributed processing networks can eliminate the need for rules and for rule induction mechanisms in the explanation of human language is unwarranted. In particular, we argue that the shortcomings are in many cases due to central features of connectionist ideology and irremediable: or if remediable. only by copying tenets of the maligned symbolic theory.
>
> (Pinker and Prince, 1988, p.82)

But it was just the first battle. Despite – or possible because of – the harsh rebuttal by Pinker, artificial neural networks suddenly met significant success in linguistics, especially among developmental linguists.

Soon many new neural models of the English past tense appeared, overcoming the limitation of the first model, fully exploiting the potential of the PDP project. MacWhinney and Leinbach (1991) presented a feedforward network implementation, with two hidden levels of 200 units each, learning with backpropagation. The transformation from phonological form was improved, abandoning the *Wickelfeature*, still ending in a fixed vector representing a word with 214 dimensions. The model succeeded in solving most of the problems that Pinker and Prince characterized as irremediable. The English past tense continued to attract the attention of those cognitivists persuaded by the new empiricist turn, and further models were implemented, each one progressing some aspect from the early model of Rumelhart and McClelland (Daugherty and Seidenberg, 1992; Plunkett and Marchman, 1993, 1996; Westermann, 1998; Plunkett and Juola, 1999). A notable example of insights into the English past tense phenomenon revealed by neural models is the blurring of regularity and exceptions. It is the tendency of many irregular verbs to exhibit aspects of the regular pattern. There are various classes of patterns of the regular forms inherited in exceptions, like the ending in /d/, as in *said*, *did* and *told*. This important aspect is neglected in the rationalist account that posits two separate mechanisms governing the past tense, a lexical one for exception, and a rule-based one for regular verbs. By contrast, PDP models capture the aspects of regularity in the exceptions because there is a unique neural network processing all verbs.

Meanwhile, Pinker built his intellectual agenda by waging war against neural and empiricist approaches to the English past tense (Kim et al., 1991; Pinker, 1991; Prasada and Pinker, 1993; Pinker, 1999, 2001; Pinker and Ullman, 2002b,a), along with the campaign against empiricism in the broader sense (Pinker, 1994, 1997, 2002). He also steered towards the same topic, one of his most brilliant students, Gary Marcus, who soon contributed in challenging the empiricist approach (Marcus, 1995). On the opposite side there are those who think the adventure into the modeling of the English past tense has had an impact far beyond the limited scope of the subject of investigation. It demonstrated the effectiveness of applying the new instruments of empiricist AI to substantive problems in cognitive science (McClelland, 1988; Plunkett and Marchman, 1996; MacWhinney, 1998; McClelland and Patterson, 2002; Westermann and Plunkett, 2007).

4.2.2   Neural Networks that Remember

Following the empiricist wave in AI, despite significant results, the issue of representing words with neural layers in the effort to model the English past tense remained largely unsatisfactory. If it was already a strain to force a single word to be encoded in a fixed-size vector, this strategy certainly became impractical by moving from a single word to a whole sentence. This difficulty would have precluded any possibility for the venturing of neural networks into the realm of syntax. Moreover, it was certainly embarrassing not being able to model such a fundamental format in cognition. Serial order is necessary for language, but it is also important for basic cognitive tasks such as goal-directed behavior, planning, or causation. How the brain represents serial order has been a crucial question, raised, for example, by Lashley (1951), and one of the most accepted solutions is

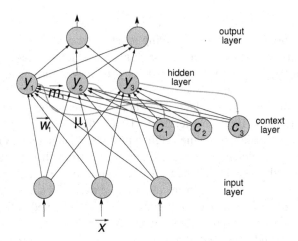

Figure 4.5: Sketch of Elman's neural recurrent network. The mathematical symbols correspond to those used in Equations (4.11) and (4.12) in the text.

the theory of short-term memory or working memory (Baddeley, 1967; Baddeley and Hitch, 1974; Baddeley, 1992).

An elegant solution was proposed by the psycholinguist Jeffrey Elman (1990) with a scheme often referred to as the 'Elman Network', is sketched in Figure 4.5. It is based on an earlier idea by Jordan (1986) of feedforward networks with added recurrent connections. As shown in Figure 4.5, the hidden layer is supplemented by a layer with the same number of units, called *context layer*. The activation of the units in these layers is computed as following:

$$y_i^{(t)} = h\left(\vec{w}_i \vec{x}^{(t)} + \vec{m}_i \vec{c}^{(t)} - \theta_i\right) \tag{4.11}$$

$$c_j^{(t)} = \begin{cases} 0.5 & \text{if } t = 0 \\ \mu_j y_j^{(t-1)} & \text{if } t > 0 \end{cases} \tag{4.12}$$

where y_i is a unit in the hidden layer, c_j a unit of the context layer, and $\vec{x}$ is the vector of inputs. Equation (4.11) is very similar to (4.4) for hidden layers in feedforward networks, and the meanings of the connection vector $\vec{w}_i$, the threshold θ_i, and the activation function $h(\cdot)$ are the same. In addition, there is now a vector $\vec{m}_i$ of synaptic connections from the context layer to the unit i in the hidden layer. Most of all, activation now depends on the time t. The account of 'time' here is not that of continuous physical time, but of discrete events of a sequence, which can be a temporal sequence of any type of ordered sequence. Therefore, $t \in \mathbb{N}$, and $t = 0$ at the beginning of the sequence. Equation (4.12) rules the activation of a unit c_j in the context layer, which is just a copy of the hidden layer at the previous time step, modulated by the connection weight μ_j, often equal to 1. At the beginning of the sequence, a constant value is assumed for the context: 0.5 in the original network of Elman. As long as the sequence progresses in time, the hidden layer receives both the current input, and the 'context' that keeps memory of what the hidden layer has received in the past.

Elman performed several experiments with its recurrent model, including abstract logical sequences, letter sequence discovery, segmentation of words inside sentences. We describe here in some detail only the most significant experiment, the one that questioned the famous autonomy of the syntax advocated by Chomsky. As described in §3.2 in generative account, syntax is the abstract structure that fixes the role of words in the sentences, constraining their order, operating at a level largely independent of lexical semantics, and higher than the surface order of words. The point made by Elman is that, whatever the abstract underlying structure of a sentence be, the surface order of words is the only information available to the listener, and words inextricably link their order to their meaning.

The experiment exposed the model to a small vocabulary made of 29 nouns and verbs, arranged in ten thousand simple plausible sentences, like `girl eat bread` or `man break car`. The input and output layers are vectors of dimension 31, therefore able to represent each lexical item as orthogonal vectors with just one dimension 1 and all the others 0. The hidden layer has 150 nodes, so does the context layer. The network is trained by backpropagation with the task of predicting the next word in the sentence, the previous words having already been input.

Note that in this case, the expected result is not in the output of the network. There is no way to predict exactly, the next word in a sentence. However, in order to carry out the prediction task, the network developed an internal representation of the lexical items, where syntactic and semantic features are combined. The internal representations are instantiated as activation patterns, in response to each word, in the hidden units. Words are represented in a space of 150 dimensions (the number of units in the hidden layer). Elman used standard hierarchical clustering analysis to analyze the organization of this space. First, the mean vector in the hidden layer is computed for each word, averaging over all occurrences of each word in all possible contexts, Then, vectors are compared for similarity, and the more similar grouped in binary hierarchies. The resulting tree is shown in Figure 4.6. The model has demonstrated that words are first organized in two main categories: verbs (at the top in the figure) and nouns (in the bottom of the figure). Verbs are further divided into groups: those which require a direct object; verbs which are intransitive; and those for which a direct object is optional. There are two major subcategories in the noun category: animate and inanimate. Animates are further divided into human and non-human; the non-humans are subdivided into large animals and small animals.

This simulation has revealed several other unexpected features. For example if, instead of averaging over contexts of a word, clustering is done separately for each context, the overall structure is similar to the tree in Figure 4.6, but each terminal leaf is now replaced with further arborization. This finding bears on the well known type/token problem: how can the brain represent, at the same time,

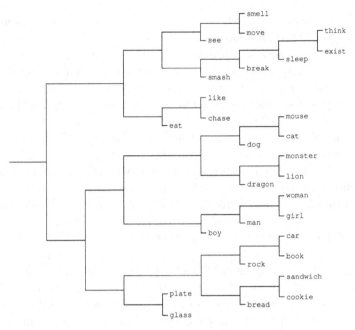

Figure 4.6: Hierarchical clustering analysis of the hidden layer in the experiment of Elman. The average vector over all possible contexts has been computed for each word.

a type, for example boy, and maintain multiple representations for more than one boy in the discourse, tokens of boy. In the representation space of Elman's network, tokens have representations which are extremely close in space, closer to each other by far than to any other types. Moreover, the organization within the token space reflects differences in the role assumed by the tokens, which are also found among tokens of other types.

Elman's network paved the way for a number of other empiricist syntax learning proposals based on recurrent neural models (Elman, 1991, 1992, 1993, 1995; Miikkulainen and Dyer, 1991; Miikkulainen, 1993; MacWhinney, 1994; Morris and Cottrell, 1999).

4.2.3 Distributed Semantics

Rumelhart and McClelland (1986a), with their model of the English past tense formation, aimed at simulating a phase in language acquisition, replicating all the main features found in psycholinguistic studies on children. Elman's aim (1990) was different, his target was not a specific pattern of language development in children; rather, it was the principle of syntax acquisition from exposure to the regularity of a corpus of sentences. McClelland completed the neural models' coverage of the empiricist perspective of language, venturing into semantics, and again, taking developmental psychology as a reference.

The forerunner of this adventure was Hinton (1981), who attempted to capture with neural networks semantic networks, the cognitive scaffolding of concepts theorized by Ross Quillian (1967, 1968). Semantic networks are made of nodes, corresponding to words, interconnected by different kinds of associative links, and arranged hierarchically. According to Quillian, the design principle of human semantic memory is well described by semantic networks. The crucial limitation of Hinton's model was that, at the time, he had not invented backpropagation (see §4.1.2). Therefore, he had to handcraft the distributed representations in his network to illustrate the relation with the idea of semantic networks.

Later on, when backpropagation made learning in PDP-like neural networks effective, Rumelhart (1990); Rumelhart and Todd (1993) progressed the model of Hinton, showing its ability to capture interesting features of Quillian's networks. In the meantime, new evidences were found within developmental psychology, notably by Frank Keil (1989, 1991, 1994) and Jean Mandler (1988, 1992), on the gradual construction of semantic networks in children, starting from broad, global concepts. In the early models of Rumelhart, McClelland envisaged a starting point for an empiricist theory of semantic cognition, where structured knowledge representation is gradually learned. Moreover, he conceived the possibility that the same neural model that learns semantics, may show its gradual reversal, simulating the degradation of semantic knowledge. McClelland would have wanted Rumelhart by his side in this new scientific enterprise but, thanks to a staggering irony, this was impossible, as he reports:

> While working with Rumelhart's model in that context, McClelland learned of the work of John Hodges, Karalyn Patterson, and their colleagues, and of the earlier work by Elisabeth Warrington on the progressive disintegration of conceptual knowledge in semantic dementia. At that time the two of us (Rogers and McClelland) began to work together on the extension of the model to address these phenomena. We hoped that Rumelhart would also participate in these investigations. Tragically, this became impossible, since Rumelhart fell victim to this very condition himself.
>
> (Rogers and McClelland, 2006, p.xii)

Having sadly lost Rumelhart's support, the project was carried on by McClellandi and Roger (2003); Rogers and McClelland (2006). The basic neural model used in all experiments of semantics is depicted in Figure 4.7. The input to the network is a pair formed by a lexical item and a relation, as it appears in a sentence. The output of the network is a layer of attributes. There are eight items, four plants such as `pine` or `rose` and four animals, like `robin` and `sunfish`. There are four possible relations: `isa`, `is`, `can`, `has`. Each relation specifies the role of the context in which the item is encountered, for example `isa` specifies a categorization context like in `daisy is a flower`, while `is` corresponds to a context in which the appearance properties of the item are stated, as in, `canary is yellow and pretty`.

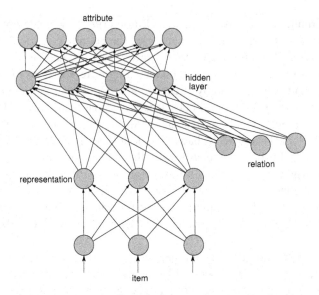

Figure 4.7: Sketch of the PDP model used for simulating distributed semantics.

The task of the neural network is to learn the attributes pertaining to the lexical item according to a relation. After learning, the network should switch on all the correct attributes for a pair item/relation, and leave off all the others. Here are examples of input pairs and expected attributes:

item	relation	attributes
robin	has	wings feathers
oak	is	living tall green
salmon	can	grow move swim

When the model has reached a reliable accuracy in predicting attributes for all item/relation pairs in the training set, semantics about the items is captured in distributed form by the representation and the hidden layers. Rogers and McClelland analyzed the progressive differentiation that semantic representation undergoes during learning. First, there is a gradual divergence in the activation of units in the representation layer for each of the eight items. At the beginning, the patterns of activation were similar for the items; at the later stage the patterns corresponding to various animal instances are similar to one another, but are distinct from those for the plants. At the final stage each item has a unique representation, but items from the same intermediate cluster, such as sunfish and salmon have similar patterns.

In the final stage, the model can learn new items, and make inferences about as yet unknown properties of those items. For example, by learning the association sparrow/isa with bird, the model is then able to activate, in reaction to the input sparrow/can, the attributes fly, move, grow, and sing. The semantic competence acquired by the model allows it to activate output units corresponding to shared characteristics of the other birds.

Other experiments with the model, using an augmented set of 21 items, addressed properties of children's early naming behavior, such as the initial learning of names that describe objects at the basic level rather than superordinate categories, or over-extension of early names (like dog) to other objects of the same category Mandler et al. (1991). The results of the experiment still shows progressive differentiation, like in the previous model, beginning with the basic distinction between animals and plants, and progressing to finer differentiation. Results indicate earlier mastery of the frequently encountered basic level names than that of more general or more specific names, with clear effects of overextention of the most frequently encountered names.

Rogers and McClelland also investigated the opposite of the evolution of semantic competence: degradation. It was simulated by adding noise to the inputs of the representation level, and the effects resemble the loss of neurons in semantic dementia. Increasing levels of noise degrade the network's ability, first to activate specific features of the item, and later to general properties of the category of the item. In addition, attributes that an item does not have, but which are typical of its superordinate domain, are applied to it when they should not. All these effects are typical of the destruction of neurons involved in representation of concepts in the brain (Warrington, 1975; Warrington and Shallice, 1984).

While the exploration of the semantic side of language by Rogers and McClelland did not spark a war comparable to that on the English past tense, it still challenged the mainstream nativist position about the nature of concepts. In philosophy, this view was championed by Fodor (1983) with the speculation on the modularity of mind. Nativism of concepts has found strong support in psychology with the theory of *core knowledge*, a set of unobservable characteristics shared by different entities, that gives rise to the observable attributes and behaviors, and glues entities as being the same kind of thing. For example, the concept dog might include such core properties as agency and rationality. Carey and Spelke (1994, 1996) emphasize that some important core properties cannot be acquired through experience and should be innate. This view is embraced and highlighted by Pinker (2002). The experiments of Rogers and McClelland include attributes that belong to the *core knowledge*, suggesting however, that these properties guide the learning of concepts because of their coherent covariation with other properties across experiences of objects.

The neural network of Rogers and McClelland was a milestone in the empiricist account of distributed semantics, but it was certainly not the only one. It could, however, be considered, the pioneer of a series of explorations of semantics with PDP-like models (Cottrell and Plunkett, 1994; MacWhinney, 1994; Regier, 1995; Dorffner et al., 1996; Cangelosi and Parisi, 1998).

In summary, the empiricist turn in AI had a profound impact on cognitive science in the domain of language, challenging its rationalist foundations. The title of a book that reflects on the first years of this movement, written together by psychologists, linguists, and members of the PDP group (Elman et al., 1996), is emblematic: *Rethinking Innateness*.

4.3 Autopoiesis and Self Organizing Systems

There is a source of theoretical inspiration for AI that does not have, even remotely, the solid foundations of empiricism or rationalism; however, it has had some fortune with AI and cognitive science as well. Most of the influence in AI derives from the idea of *self-organization*, a notion that recurs in a variety of scientific fields. In the context of relevance for AI and cognitive science, a philosophical ground for self-organization is given by the concept of *autopoiesis*.

4.3.1 The Philosophy of Autopoiesis

The term *autopoiesis* was proposed by the Chilean biologists, Humberto Maturana and Francisco Varela (1980), as the exclusive property that can define a living being: "we claim that the notion of autopoiesis is necessary and sufficient to characterize the organization of living systems" (p.82). For Maturana and Varela, all living systems are *autopoietic machines*, defined as follows:

> An autopoietic machine is a machine organized (defined as a unity) as a network of processes of production (transformation and destruction) of components that produces the components which: (i) through their interactions and transformations continuously regenerate and realize the network of processes (relations) that produced them; and (ii) constitute it (the machine) as a concrete unity in the space in which they (the components) exist by specifying the topological domain of its realization as such a network.
>
> (Maturana and Varela, 1980, p.78–79)

It may be needless to say that the search for a definition of life is a core question in biology; according to Erwin Schrödinger (1944) "What is life?" is the mother of all scientific questions. The answer proposed by Maturana and Varela was revolutionary for reducing all that is life to a single abstract property: autopoiesis. By looking – for example – at the fifty or more different definitions of life collected by (Popa, 2004, p.197–205), most of them consist of a combination of several properties that, together, contribute to life. In fact, the two characteristics spelled in the above description of an autopoietic machine bear resemblance with other properties used by biologist, in particular (ii) corresponds to the physical separation of a living system associated with its boundaries. The characteristic (i) resembles the (M,R) systems (*metabolism-repair*) developed by Robert Rosen (1958, 1991), but the autopoietic account abstracts away from all physical aspects of metabolism. The theory of autopoiesis emphasizes the circular closure of living beings: their network of processes reaches a stable organization by the interaction between its components and the interactions that they, together, maintain with the external of the organism. In other words, the organization produces an activity that is both cause and effect of the very organization: *self-organization.*

From a strictly biological point of view, autopoiesis neglects important aspects associated with life, such as bioenergy (dissipation, irreversibility, coupling, energy currencies) (Popa, 2004, p.15) and regulatory mechanisms for metabolism (Cornish-Bowden and Cárdenas, 2020, p.33). Other aspects, such as reproduction and evolution, that in orthodox biology are deemed essential for life (Maynard Smith and Szathmáry, 2000), are secondary in autopoiesis. In fact autopoiesis has had a marginal impact on mainstream biology but, thanks to its generality, it enjoyed widespread influence for some time. Probably the most enthusiastic advocate of autopoiesis was the German sociologist Niklas Luhmann (1984), who described social systems as autopoietic networks of processes reproduced by communication. More compelling for AI is the close link made by Maturana and Varela between autopoiesis and cognition, in their own words:

> A cognitive system is a system whose organization defines a domain of interactions in which it can act with relevance to the maintenance of itself, and the process of cognition is the actual (inductive) acting or behaving in this domain. Living systems are cognitive systems, and living as a process is a process of cognition. This statement is valid for all organisms, with and without a nervous system.
>
> (Maturana and Varela, 1980, p.13)

The perfect equivalence between living systems and the cognitive systems, since they are autopoietic categories, is decidedly perplexing. As observed by Margaret Boden (2000, p.139), "it is unnecessary and confusing to widen the scope of cognition so as to include all living things – including algae and flowering plants." The ascription of cognition within the autopoiesis account becomes so liberal as to lead Dennis Dollens (2014) to seriously rephrase Turing by asking: "Can buildings think?"

The contribution of autopoiesis inside cognition is characterized by the rejection of informational concepts such as representations; and by sympathies towards phenomenology and embodied cognition (Varela et al., 1991). Ultimately, autopoiesis has found favor with those who, as noted in §1.4, find the computational perspective of mind disturbing, and are attracted to the alternative scenarios suggested by Varela such as the trendy Buddhist method of *mindfulness meditation*.

4.3.2 Self-Organization made Computational

Self-organization in the context of autopoiesis is a valuable proposal for theoretical biology and, to some extent, for cognition. However, it would have no relevance for AI, without an appropriate set of computational suggestions, suggestions that could hardly have come from the philosophy of Maturana and Varela, in which the concept of computation had no place. Early proposals for a computational account of self-organization arrived instead from other fields like physics, chemistry, cybernetics and theoretical neuroscience. The former owes its name to Norbert Wiener (1948), who defined it as the study of 'control and communication in the animal and the machine'. Cybernetics had a remarkable success, especially in the States and in UK, after the Second Wold War (Heims, 1991), and it constituted a fundamental precursor for both AI (Cordeschi, 2002) and cognitive science (Dupuy, 1994).

Ross Ashby (1947) was the first to use the term 'self-organization', making this concept central for cybernetics. He first gave a broad mathematical definition of a self-organizing system. The status of such a system, at a given moment, should be captured by a set of variables $\{x_1, \cdots, x_n\}$, and the behaviour of the system in time may be specified by equations of the form:

$$\frac{dx_i}{dt} = f_i(x_1, \cdots, x_n). \tag{4.13}$$

The 'organization' of the system is the set of functional forms $f_i(\cdot)$ in Equation (4.13). The system is able to self-organize if one of the variables appears as a step-function of time. Let x_n be the step-function, and let us assume for simplicity that it has two values only:

$$x_n = \begin{cases} X' & \text{for } t < \tau \\ X'' & \text{for } t > \tau \end{cases} \tag{4.14}$$

where τ is the time instant when x_n changes. As x_n is a constant when $t < \tau$, it may be absorbed into a new functional that takes its value X' into account. After x_n has changed to X'' it remains constant and we can again absorb it into a new functional. We have now two 'organized' systems:

$$\frac{dx_i}{dt} = \begin{cases} g_i(x_1, \cdots, x_{n-1}) & \text{for } t < \tau \\ h_i(x_1, \cdots, x_{n-1}) & \text{for } t > \tau \end{cases} \tag{4.15}$$

where $g_i(\cdot)$ and $h_i(\cdot)$ are the functional forms including the constant values, respectively, X' and X''.

A concern for Ashby was to reconcile two apparently clashing facts found in biological systems or artificial machines, that of being strictly determined in their actions by physical-chemical processes, and yet able to undergo self-induced internal reorganizations resulting in changes of behavior. In one of his later papers, Ashby (1962) related the idea of 'organization' to the complexity of the system, such as the dependence of the system's behavior upon a usually high number of interacting variables. It contrasts with a system in which variables can be separated in mathematical forms. For a system to develop a certain organization under this definition is rather trivial, and does not necessarily qualify as a case of 'self-organization', which, for Ashby, is the property of changing from a 'bad' to a 'good' organization. However, there is no a priori criteria for evaluating the developed organization as being good in any absolute sense; not only does it depend on each specific system, it is also a property that is observable only from the outside, that is, only when the organization has been reached. In the case of a brain, for instance, an organization can be deemed 'good' if it acts so as to provide some kind of advantage to the organism's survival.

Ashby, unlike Maturana and Varela, and on the same lines of AI, holds that the computer is the ideal tool for the creation and understanding of intelligent – self-organizing – systems:

> [...] every dynamic system generates its own form of intelligent life, is self-organizing [...] Why we have failed to recognize this fact is that until recently we have had no experience of systems of medium complexity; either they have been like the watch and the pendulum, and we have found their properties few and trivial, or they have been like the dog and the human being, and we have found their properties so rich and remarkable that we have thought them supernatural. Only in the last few years has the general-purpose computer given us a system rich enough to be interesting yet still simple enough to be understandable.
>
> (Ashby, 1962, p.270)

Actually even before the birth of computers a few suitable examples of observable self-organization phenomena had been found. A striking case was observed as early as the beginning of the last century by the French physicist Henri Claude Bénard (1900), that of convective cells that organize in a fluid. When heating water in a pot, buoyancy produces the upwelling of lesser dense molecules, which for conservation of mass, should be compensated by the downward motion of colder molecules. While initially the billions of water molecules exhibit a random motion, gradually, a small number of regular patterns of cells organizes. The cells are formed by an inner cylinder with a laminar upward flow, bounded by downward flow at the periphery. We usually miss this amazing phenomena because water is transparent, it can be experimentally visualized using solid markers, as in the left image of Figure 4.8.

Figure 4.8: Examples of self-organization, natural and man-made. On the left, a case of Bénard's convective cells, with heated silicon oil, the cells made visible thanks to a graphite marker. In the middle, the B-Z reaction. On the right, Ashby's Homeostat machine.

Something very similar happens in chemical reactions. The B-Z reaction, for example, named by its discoverers Belousov (1959) and Zhabotinsky (1964), is well known. It is a mix of potassium bromate, cerium(IV) sulfate, malonic acid and citric acid in dilute sulfuric acid. The malonic acid reduces cerium(IV) ions into cerium(III), which in turn tends to be oxidized back to cerium(IV) ions by the potassium bromate. Here again, the alternate disposition of the two ion types of cerium is initially random, but gradually tends to organize into macroscopic patterns, like those shown in the middle in Figure 4.8.

Ashby actually engaged himself in the construction of an artificial self-organizing system, named *Homeostat* (Ashby, 1949), shown in the right in Figure 4.8. It was made of four units, each carrying on top a suspended magnet. The end of the suspending wire extends into the water in a plastic trough which has electrodes at each end. The potential picked up by the suspending wire depends, at each moment, on the position of the magnet. In turn, each magnet is affected by the currents in the three coils of the other units, and partly by its own coil. This arrangement sets all four units into action and reaction on one another: the magnets are moved by the currents, but these movements change the currents, causing fresh movements.

So far, the behaviour of the system has been based on feedback control, which is the key engineering notion exploited by cybernetics. Ashby's Homeostat is a step further, because the amount of feedback is not set by hand, it can be switched by an automatic uniselector in each unit. There are 25 different feedback regulations for each unit, and the regulations are changed whenever the system becomes unstable, i.e., whenever some magnet diverges too far from the central position. The surpring result is that, even if the 25 feedback settings for all units are random, the machine hunts for a combination of uniselectors that ensure a proper internal feedback. Every change in uniselectors gives rise to a newly 'organized' machine. When a suitable combination is found, it is held. In other words, the machine has self-organized.

In the same period, a mathematical description of self-organization in the brain was pioneered by the neuroscientist Raymond Beurle (1956, 1962). He addressed systems made by a large number of interacting units, quite like molecules in heated water or in chemical reactions. The real interest is in the behavior of neurons, but for simplicity he just assumed simpler cells, able to excite other connected cells. Beurle studied waves of activations of randomly connected cells, identifying properties such as attenuation or gain in wave amplitude, saturation, dependency on density of connections. The approach made use of analogy with the well established field of wave analysis in physics. A different strategy was suggested by Farley and Clark (1954), for the same purpose of studying systems made by a large number of interacting simplified neurons, but simulating individual discrete units.

An important step moving forward was a series of works by the neuroscientists, Christoph von der Malsburg and David Willshaw, trying to explain certain aspects of the visual system by self-organization. In the first work von der Malsburg (1973) was concerned with one of the most striking features of the visual cortex: the ordinate arrangement of neurons responding to light bars and edges of a certain orientation (Hubel and Wiesel, 1963, 1968). A model was developed, made of 338 neurons arranged in two dimentions as in the cortex, connected to a retina-like source where oriented stimuli were applied. After several retinal experiences, individual neurons became sensitive to only one orientation, and clusdered in a way analogous to the mature cortical organization. Other works Willshaw and von der Malsburg (1976); von der Malsburg and Willshaw (1977) addressed the topographically ordered mapping, where neighbouring distal cells project to neighbouring cortical neurons. This kind of organization is fundamental in primary sensorial corteces (Mountcastle, 1957; Hubel and Wiesel, 1974), but is also found in non mammal animals (Cooper et al., 1953; Gaze, 1958), and von der Malsburg and Willshaw demonstrated how it can be established by self-organization. In (von der Malsburg and Willshaw, 1976; von der Malsburg, 1979) a different phenomenon of the visual system is explored: the formation of ocular dominance strips. This is the case in animals with binocular vision, in which connections originating in the two eyes innervate a target structure (the primary cortex, the optic tectum, or the thalamus). Ocularity domains are formed by the fibres of one eye concentrating in small domains of the target structure, leaving the remaining space to the fibers of the other eye (Hubel and Wiesel, 1962, 1970).

All the above models share a number of assumptions and a common mathematical basis for the self-organization of the aspect of vision under investigation. There are three key mechanisms in cortical circuits that match with the premises of self-organization:

1. Small signal fluctuations might be amplified, this is a direct effect of the non-linear behavior of neurons;

2. there is cooperation between fluctuations, in that excitatory lateral connections tend to favor the firing of other connected neurons, and Hebbian law reinforces synapses of neurons that fire frequently in synchrony;

3. there is competition as well, in that inhibitory connections can lower the firing rate of groups of cells at the periphery of a dominant active group, and synaptic homeostasis compensates for the gain in contribution from more active cells by lowering the synaptic efficiency of other afferent cells.

In all models, the activity x_i of a neuron i in the target area is computed by systems of differential equations that can be generalized as follows:

$$\frac{\partial}{\partial t} x_i(t) = -\alpha_i x_i(t) + \sum_{j \in \mathscr{C}_i} w_{ij} f(x_j(t)) + \sum_{j \in \mathscr{A}_i} w_{ij} a_j(t) \qquad (4.16)$$

$$f(x_i(t)) = \begin{cases} x_i(t) - \theta_i & \text{if } x_i(t) > \theta_i \\ 0 & \text{otherwise} \end{cases} \qquad (4.17)$$

where $\mathscr{C}_i$ is the set of neurons in the target area with lateral connections to the cell i, and $\mathscr{A}_i$ is the set of all afferent axons, each carrying a signal $a(t)$. w_{ij} is the synaptic efficiency between cell presynaptic j and postsynaptic i, and is modified by an amount proportional to the presynaptic and postsynaptic signals in the case of coincidence of activity. Periodically, all w_{ij} leading to the same neural cell i are renormalized, realizing the competition, in that some synapses are increased at the expense of others. The target area can be the primary visual cortex in the case of mammals, or the optic tectum in other vertebrates. The source of afferents, leading to the process of self-organization, can be not only the external scene seen by the eyes, but also spontaneous activity generated by the brain itself.

4.3.3   Self-Organization made Science

Nowadays laser devices have become ubiquitous; they are common in discos, widely used in consumer electronics, medicine, industry, and even applied for headlamps on luxury cars. At the time of its invention by Theodore Maiman (1960), the theoretical principles behind the operation of the laser were obscure and highly intriguing. One of the most advanced institutes for laser theory in Europe established by the physicist Hermann Haken in Stuttgart. His studies lead to an interpretation of the physics of the laser as a self-organization phenomenon (Haken, 1964a,b, 1965, 1969).

Here, in a later book by him for the more casual reader, is a non-mathematical description of what happens in a laser:

> When we start pumping energy into the laser, the following happens: At small pump power the laser operates as a lamp. The atomic antennas emit independently of each other, (i.e., randomly) light-wavetracks. At a certain pump power, called laser threshold, a completely new phenomenon occurs. An unknown demon seems to let the atomic antennas oscillate in phase. They emit now a single giant wavetrack whose length is, say 300,000 km!
>
> (Haken, 1978, p.5)

The laser is a system lying on the borderline between a natural system and a man-made device, but Haken was well aware that 'unknown demons' driving systems into powerful coherent organizations are widespread in nature. Therefore, he engaged in the research program of unveiling the theoretical principles behind the unknown demons of natural self-organization, and he called this new-born science *synergetics* (Haken, 1978, 1983, 1984). Let us provide just a brief mention of one of the pivotal equations in the framework of synergetics. It is the Fokker-Planck equation, named after the physicists Adriaan Fokker and Max Planck, aimed at solving the difficut problem of describing the behavior of inter-acting units, when their number is too large. In cases like the convective cells of Bènard the number of units is of the order 10^24, and it is impossible to solve a system of coupled equations, each describing the exact motion of an individual particle. The only possibility is to use stochastic quantities for the properties of interest of the units. The Fokker-Planck has several formulations depending on the specific system it applies to (Risken, 1989). The following is an example:

$$\frac{\partial}{\partial t}p(\vec{q},t) = -\nabla_{\vec{q}}p(\vec{q},t)\vec{k} + Q\Delta_{\vec{q}}p(\vec{q},t). \tag{4.18}$$

It describes the evolution in time of the probability density p for a point $\vec{q}$ in a space, for example the 3 geometrical coordinates of the position of a molecule, where k is a velocity field in the space of $\vec{q}$, and Q is the correlation function of the random fluctuations of $\vec{q}$. This equation statistically links the microscopic level of the motion, or other general characteristics of the elementary components of a system, with its mesoscopic level.

In the early years, the main focus of synergetics was on physical and chemical systems, and marginally biological systems. Haken then gradually shifted his attention from physics to cognition and neuroscience (Haken, 1991, 1996, 2002).

The same Fokker-Planck equation that had been crucial for Haken (1969) in describing the behavior of the laser can be used for cognitive tasks like visual recognition (Haken, 1991, Ch.5). The vector $\vec{q}$ in this case is made by pixels in an image, and the evolution in time is from an input image to a prototype pattern $\vec{v}_j$, representative of the j known visual category of the agent. The trajectory in time of vector $\vec{q}$ would be:

$$\vec{q}(0) \rightarrow \vec{q}(t) \rightarrow \vec{v}_j \tag{4.19}$$

where the full trajectory can be derived with Equation (4.18) (not the easiest job to do), and $\vec{v}_j$ is a kind of basin of attraction for $\vec{q}(0)$.

4.3.4 *Self-Organization made Easy*

Despite Haken's latest focus on cognition, synergetics has had a marginal impact on cognitive science, and even less on AI. One reason was definitely the lack of ease of use of the methods of sociology, whose mathematics was very familiar to physicists, but certainly not to psychologists. The success of artificial neural networks in cognitive science and linguistics, described in §4.2, was also assured by the extreme simplicity of the connectionist tools. In the same period as the boom in neural computation after the PDP project of Rumelhart and McClelland (1986a), a proposal arrived, offering self-organizing properties at a much cheaper price than synergetics, with the same simplicity and efficiency of the other models in the connectionist arena.

This algorithm is known with the acronym SOM (*Self-Organizing features Map*), and is thanks to the Finnish electronic engineer Teuvo Kohonen. He was originally interested in psychology in the context of *associative memories* (Kohonen, 1977), a strategy in computer science loosely inspired by human memories. The basic operation of associative memory is the storage of information together with the relations or links between the data items, and the selective recall of stored information relative to a piece of releted information presented. In this period, Kohonen went to German conferences on neural networks and he become acquainted with Christoph von der Malsburg (see §4.3.2). Von der Malsburg's models of self-organization in the visual system have been the main proposition for Kohonen, rather than synergetics. He convinced himself that a central issue for memories, at a more abstract level, is the self-organization of knowledge, namely, how the concepts themselves could be formed in a learning system (Kohonen, 1984). SOM, Kohonen's main achievement, was conceived as a generalization and a simplification of von der Malsburg's models:

> I just wanted an algorithm that would effectively map similar patterns (pattern vectors close to each other in the input signal space) onto contiguous locations in the output space. Ch. v.d. Malsburg had obtained his pioneering results in 1973, but I wanted to generalize and at the same time ultimately simplify his system description.
>
> (Kohonen, 1995, p.X)

The SOM algorithm fully met the expectations: self-organization made as easy as the PDP project of Rumelhart and McClelland (1986a). The acronym SOM indicates that the space where self-organization will take place is two dimentional: a *map*. Maps are also the most basic arrangements of neurons in primary sensory cortical areas (Mountcastle, 1957, 1978; Kaas, 1997), and Kohonen (1998) speculates that maps might be a fundamental organization in higher areas of the cortex too. In actual fact, the account of 'map' is more general, in that there is no specific number of dimensions, although the two-dimensional case is the most common and, clearly, the most suitable for simulating perceptual phenomena.

The SOM has its mathematical roots in *Vector Quantization*, a method used in signal processing to approximate with a small number of vectors, called *codebook*, the probability density of a stochastic high dimensional vector (Linde et al., 1980).

The mathematics of the SOM is elegant and, indeed, very simple $\vec{t} \in \mathbb{R}^N$ being the data to analyze, a two dimensional SOM is made up of M neurons $\vec{x} \in \mathbb{R}^N$, with an associated two-dimensional coordinate $\vec{r} \in \{< [0,1], [0,1] >\} \subset \mathbb{R}^2$. When data are presented to the network, the same vector is available to all neurons in the map. The main strategy of the algorithm is the so-called *winner-take-all*, the singling out of just one neuron over all M, which best responds to that specific input. A scheme of this network is provided in Figure 4.9.

In mathematical terms, for a give input $\vec{t}$ the winner neuron $\vec{x}_c$ is chosen by the following equation:

$$c = \arg \min_{i \in \{1,...,M\}} \left\{ \|\vec{t} - \vec{x}_i\| \right\} \tag{4.20}$$

where the metrics for comparing two vectors is arbitrary. The same procedure is used during the learning phase of the network. At the beginning, all neurons start with random vectors and all samples $\vec{t} \in \mathscr{T}$ are presented in random order. After the selection of a winner neuron c, in response to a sample $\vec{t}$, using (4.20), the vectors associated with the neurons are modified using this rule:

$$\Delta \vec{x}_i = \eta e^{-\frac{\|\vec{r}_c - \vec{r}_i\|^2}{2\sigma^2}} \left(\vec{t} - \vec{x}_i \right) \tag{4.21}$$

where η is a the learning rate, and σ is the width of the influence of the winner c in adapting to its neighbors. There are two components in (4.21), this one

$$\eta \left(\vec{t} - \vec{x}_i \right) \tag{4.22}$$

attracts $\vec{x}_i$ to the target $(\vec{t}$, by an amount weighted by η. Equation (4.22) coincides with (4.21) in the case of the winner, $c = i$. For all the other neurons, an additional modulation is given by the second term of (4.21):

$$e^{-\frac{\|\vec{r}_c - \vec{r}_i\|^2}{2\sigma^2}} \tag{4.23}$$

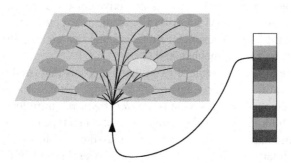

Figure 4.9: Scheme of a two dimensional Kohonen map. Every neuron receives the same vectorial input, as shown on the right. Only one neuron is the 'winner', here marked in white.

limiting the vector update, as long as the neurons are far away from the winner, with a Gaussian shaped by σ. At the end of the learning phase, a relational ordering may come about in the map, so that the presentation of a sample $\vec{i}$ will trigger a winner neuron in a spatial location of the map, relevant for some feature of the data.

The SOM algorithm has been successful in a range of AI applications, from robot control (Ritter et al., 1992; Littman et al., 1992) to combinatorial optimization (Plebe and Anile, 2001; Plebe, 2001) and document mining on the web (Lagus et al., 1999; Kohonen et al., 2000).

In addition, the SOM made self-organization popular and manageable among psychologists and linguists. Kohonen himself has undertaken some digressions in the domain of language. Ritter and Kohonen (1989) demonstrated the self-organization of categories in two experiments with a restricted lexicon.

The simplest experiment uses 16 animal nouns, each one has an associated 29-dimensional binary vector indicating features such as size, number of legs, having feathers or hooves. The SOM map encoding this fragment of semantics has 10×10 units, and its organization at the end of the training is shown in Figure 4.10. The animal names are shown on the cell location, winner when the noun is presented to the SOM. The spatial order in the map has captured the basic relationships among the animals, with three main clusters collecting birds, aggressive, and tame animals. A more complex experiment echoed the aims of Elman's model (see §4.2.2), targeting a small vocabulary comprising nouns, verbs, and adverbs. The vectors are now generated from 498 different three-word sentences, where single words of the vocabulary are embedded by their predecessor and successor. The resulting SOM exhibits an overall organization in three domains for nouns, verbs, and adverbs, with additional subdivisions reflecting aspects of meaning.

Even though its impact was certainly less relevant than that of the PDP project (see §4.2), there are few studies that model language development with the SOM (MacWhinney, 1998; Li et al., 2004). The simplicity and effectiveness of the SOM has been an important added value for AI, but its glaring departure from brain mechanisms makes it less attractive for cognitive science. In evaluatiing how close the SOM algorithm is to the real physiology of the cortex, von der Malsburg (1995) described Kohonen maps as 'an algorithmic caricature of the [self-organization] mechanism'.

Let us give a short account of two other proposals that tried to maintain the simplicity of artificial neural networks reflecting, at the same time, some aspect of the brain mechanisms. One is the *Neocognitron*, introduced by Kunihiko Fukushima (1980, 1988). It alternates layers of *S-cell* type units with *C-cell* type units, and those names are evocative of the classification in simple and complex cells in the primary visual cortex by Hubel and Wiesel (1962, 1968). The S-units act as convolution kernels, while the C-units downsample the images resulting from the convolution by spatial averaging. The crucial difference from conven-

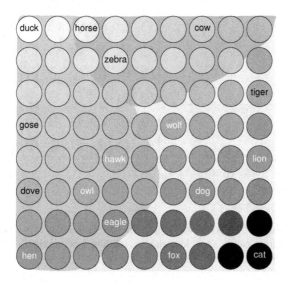

Figure 4.10: The semantic map of Ritter and Kohonen (1989) at the end of the self-organization process. The part of the map on the left in light gray is occupied by birds, the more predatory animals gather towards the right (white area), whilst the more peaceful species are concentrated in the upper middle (medium gray area).

tional convolution in image processing (Rosenfeld and Kak, 1982; Bracewell, 2003) is that the kernels are learned by self-organization. The algorithm is similar to that of the SOM with a winner-take-all strategy; only the weights of the maximum responding S units, within a certain area, are modified, together with those of neighboring cells.

A different proposal is the LISSOM (*Laterally Interconnected Synergetically Self-Organizing Map*) (Sirosh and Miikkulainen, 1997), which, as the acronym reveals, brings together concepts of self-organization, map topology, lateral connections, and synergetics. The basic equation in the LISSOM describes the activation level x_i of the i-th unit in the map:

$$x_i^{(k)} = f\left(\frac{\gamma_A}{1 + \gamma_N \vec{U} \cdot \vec{v}_{r_A,i}} \vec{a}_{r_A,i} \cdot \vec{v}_{r_A,i} + \right.$$
$$\left. \gamma_E \vec{e}_{r_E,i} \cdot \vec{x}_{r_E,i}^{(k-1)} - \gamma_I \vec{i}_{r_I,i} \cdot \vec{x}_{r_I,i}^{(k-1)}\right).$$

(4.24)

Vector $\vec{v}_{r_A,i}$ is composed by afferent to unit i in a circular radius r_A, the vectors $\mathbf{x}_{r_E,i}^{(k-1)}$ and $\mathbf{x}_{r_I,i}^{(k-1)}$ are the activation of all neurons in the map, where a lateral connection exists with neuron i of an excitatory or inhibitory type, respectively. Their fields are circular areas of radius, respectively, r_E, r_I. Vectors $\mathbf{e}_i$ and $\mathbf{i}_i$ are composed by all connection strengths of the excitatory or inhibitory neurons projecting to i. The scalars γ_X, γ_E, and γ_I, are constants modulating the overall contribution of afferents, excitatory, and inhibitory components.

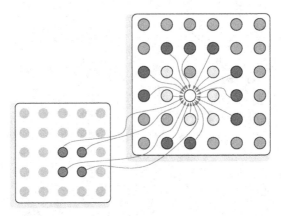

Figure 4.11: Scheme of connections of a unit in the LISSOM architecture. The unit in white color receives excitatory connections, in dark grey, inhibitory connections, in light grey, with additional afferences from thalamic maps on the left.

The scalar γ_N controls the setting of a push-pull effect in the afferent weights, allowing inhibitory effects without negative weight values. Mathematically, it represents dividing the response from the excitatory weights by the response from a uniform disc of inhibitory weights over the receptive field of neuron i. In Equation (4.24) and all those following the operation, $\mathbf{x} \cdot \mathbf{y}$ is the product of vectors $\mathbf{x}$ and $\mathbf{y}$. Vector $\mathbf{U}$ is just a vector of 1's of the same dimension of $\mathbf{x}_i$. The function f can be any monotonic nonlinear continuous growing function limited between 0 and 1; it is generally a piecewise linear approximation of the sigmoid function.

Equation (4.24) is recursive in time k, with initial condition $x_i^{(k=0)} = 0$; the final activation is defined as $x_i = x_i^{(k=K)}$, where K satisfies the following condition:

$$\sum_i \left| x_i^{(k=K)} - x_i^{(k=K-1)} \right| < \varepsilon \tag{4.25}$$

with ε a small defined value. A scheme of connections to a LISSOM unit is provided in Figure 4.11. One of the main features that make the LISSOM scheme suitable for modeling the cortex is its inclusion of computational contributions of intracortical lateral connections. In addition, there is no simplified ad hoc learning rule, as in the SOM: connections are modified by strict Hebbian learning compensated by a synaptic homeostatic mechanism.

The LISSOM architecture was originally conceived for simulating visual area V1, in direct connection with the thalamus (Bednar, 2002). It evolved into Topographica (Bednar et al., 2004), which has had several extensions, making it a valuable tool for composing complex hierarchies of cortical maps (Miikkulainen et al., 2005; Plebe and Domenella, 2007). Both Neocognitron and Topographica were mainly conceived for modeling self-organization in the visual system, but

Topographica is more abstract and general and has been also used for simulations in the domain of language (Plebe and De La Cruz, 2016).

4.4   Other Philosophical Inspirations

Empiricism and rationalism are the two main contending philosophical traditions within AI, with the less established domain of autopoiesis playing the part of the third wheel. The list is not exhaustive, other philosophical guises can be found in the variety of wisdom found in AI. In this section, we survey in brief, a few of them, each one linked to a specific great thinker of the past.

4.4.1   *Bayes*

Thomas Bayes would hardly call himself a philosopher, he was an amateur mathematician and, what is called 'Bayesian epistemology' emerged as a philosophical program only at the beginning of the last century (Earman, 1992). Bayes' rebirth in philosophy is justified by suggestions found in his main work (Bayes, 1763) for an interpretation of probabilities, together with a related mathematical framework, as mental features. Bayes' *Essay Towards Solving a Problem in the Doctrine of Chances* was published posthumously, thanks to his friend Richard Price, a moral philosopher with an interest in mathematics. Price was also engaged with statistics for professional reasons; he was consulted by Equitable, the first life insurance company. He compiled tables of mortality curves, known as Northampton tables because he used records of the city of Northampton. He dramatically underestimated mortality, setting life expectancy at birth at 24 years. In 1808 the British government issued annuities based on Northampton tables and lost millions of pounds. This failure is by no means compensated by Price's far-sightedness in understanding the value of Bayes' essay. In this work there are at least the following three ideas of philosophical relevance:

- the mental account of probability as degree of belief of people;

- the introduction of conditional probability expressing the relation between evidence and belief;

- the principle of conditionalization as a process of belief revision.

Bayes was not the first one trying to assess what probability is; a detailed historical account can be found in Hacking (1990). Let us also just mention the contributions of Jacob Bernoulli (1713) and Abraham de Moivre (1718). Bayes assumed from the last one, the expression 'Doctrine of Chances' for the study of probability, and has drawn extensively on the available studies on statistics, but the three aforementioned ideas were original and new.

Let start with the first idea. For Bayes, probabilities are degrees of confidence, or credence, or partial beliefs, held by people. For example, calling R the event that tomorrow it will rain in some place where one lives, $p(R)$ is a number between 0 and 1 corresponding to the degrees of belief she holds about raining tomorrow, with $p(R) = 1$ certainty of rain, $p(R) = 0$ certainty of no rain. This interpretation has been rejected, if not ignored altogether, in the century following Bayes. The interpretation of probability more distant from the Bayesian subjective probability, is the *frequentist* interpretation. According to frequentism probabilities are fundamentally dispositional properties of physical systems, and their value coincides with the frequency of the event, i.e., the ratio of the counts of the events to the count of all possible events. The frequentist interpretation has been advocated, among others, by John Venn (1866), Richard Von Mises (1931) and Hans Reichenbach (1949). The clear advantage of the frequentist interpretation is that it does not depend on what anyone thinks, it is an objective physical concept.

The second idea of Bayes may appear of neutral significance, as a simple definition of conditional probability:

$$p(X|Y) = \frac{p(X \wedge Y)}{p(Y)}. \tag{4.26}$$

It states that the probability of event X, given that the event Y occurs also, is the ratio between the probability that both events X and Y occur and the absolute probability of Y. It acquires, however, epistemological significance with special cases of X and Y. This is the case where X is a subjective belief, and Y is an evidence related with this belief. As Bayes highlighted, most people's beliefs are conditioned by evidences. Using the previous example, it is hard to imagine an absolute degree $p(R)$ for the belief that tomorrow it will rain, while it is more realistic to take into account a conditional probability, for example $p(R|O)$, where O is the event of an overcast sky, and $p(R|O)$ is the degree of belief that tomorrow it will rain if the sky is overcast. The epistemology related with conditional probability, known as simple principle of conditionalization, is that whatever prior absolute degree $p_0(X)$ one held for his belief X, if an event Y related with X happens, the absolute degree of the belief is updated with $p_1(X) = p(X|Y)$. So, if the individual of the example observes that the sky is overcast, his degree $p(R)$ for the belief that it will rain becomes updated to $p(R|O)$.

Eventually, the third idea of Bayes is related to the inversion of the conditional probability, appearing in his celebrated theorem:

$$p(X|Y) = \frac{p(Y|X)p(X)}{p(Y)}. \tag{4.27}$$

As in the case of conditional probability, there is nothing philosophically relevant and controversial in Equation (4.27); it becomes so with the interpretation of X as mental belief and Y as external evidence. The quantity $p(Y|X)$, dubbed the

'Bayesian heresy' by David Salsburg (2001), is rather disconcerting under the epistemic interpretation. It suggests the talking about the probability of an external event, conditioned by a mental belief, in the above example, the probability of the sky being overcast, given my degree of belief in rain. Again, since this inverse probability shackled the standard use of statistics in scientific investigation; it was, therefore, carefully avoided by statisticians. Renowned statisticians like Ronald Fisher and Jerzy Neyman had to defend their work from the accusation of using the inverse conditional probability.

The mental account of probabilities found a resurgence in the last century, supported by the works of Frank Ramsey (1931), Leonard Savage (1954), and Bruno de Finetti (1974). One of the most compelling arguments developed by these authors, and already sketched by Bayes, is known as the *Dutch book* argument[1]. It is a mathematical demonstration constructed on gambling on events for which probabilities are known. The strategy is to show that if a gambler uses a set of betting quotients that fails to satisfy the mathematics of Bayesian probability, then there is a set of bets with those quotients that guarantees a net loss, no matter the outcomes of the game. Therefore, the mathematics of Bayesian probability is the ground of our reasoning with probabilities. There are, of course, several objections to the arguments, and to its conclusions. The whole debate between Bayesianism and frequentialism is still alive, with several additional interpretations of probabilities between the two extremes but, since decades, the Bayesian position tends to prevail (Hacking, 1975, 1990; Salsburg, 2001).

It was, therefore, time that the AI community also became aware of Bayes, and the first to do so was Judea Pearl (1986, 1988), with the invention of a model he called *Bayesian networks*. Pearl erected his construction on two pillars: Bayes' probability and graph theory. Graphs are mathematical objects made of discrete nodes connected by arcs, first pioneered by Leonhard Euler (1736) to solve the *Königsberg bridge problem*. The river Pregel crosses Königsberg forming two islands connected to the banks and each other by seven bridges. The problem is to start from any land area, walk over all the bridges exactly once, and return to the starting point. By using a graph where nodes represent land areas and arcs represent bridges, Euler demonstrated that there is no solution to the Königsberg bridge problem. Since Euler, graphs become a subject of intense studies (Sylvester, 1878; Harary, 1969). The availability of a sound mathematical framework for handling graphs made them the most suitable data structure in computer science (Deo, 1974; Even, 1979).

Therefore, Pearl worked on very solid pillars to construct his Bayesian networks, where nodes represent random variables, and arcs the conditional prob-

---

[1]There is no certain origin of the term *Dutch book*; it is seemingly common among horse race bookmakers, meaning a bettor that uses such wrong rates at which he will bet, that she will lose money no matter what happens. Probably 'Dutch' is used in a derogatory sense as it often was in English, dating from the 17th century, when Holland and England had rivalry between their fleets.

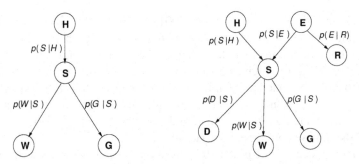

Figure 4.12: Example of Bayesian networks. In the left graph the nodes are: *H*=burglary hypothesis; *S*=alarm sound; *G*=Mrs. Gibbon's testimony; *W*=Dr. Watson's testimony. The graph on the right has the same node as that of the left with the addition of the following: *H*=burglary hypothesis; *E*=earthquake hypothesis; *R*=news report of the earthquake; *D*=daughter will call.

abilities between variables. We spare the reader the mathematics of Bayesian networks, rather we illustrate it with one of Pearl's examples:

> Mr. Holmes receives a telephone call from his neighbor Dr. Watson, who states that he hears the sound of a burglar alarm from the direction of Mr. Holmes's house. While preparing to rush home, Mr. Holmes recalls that Dr. Watson is known to be a tasteless practical joker, and he decides to first call another neighbor, Mrs. Gibbon, who, despite occasional drinking problems, is far more reliable.

> (Pearl, 1988, p.42)

In Figure 4.12, on the left, is the simple Bayesian graph representing the alarm scenario. The evidence variable *S*, the sound of the alarm, is uncertain, and even if true, it enforces but does not grant the hypothesis *H* of a burglary. Moreover, the only possible available evidences are the testimony of Watson or Gibbon, both probabilistically related to *S*. By applying Bayes' theorem, the probability of burglary *H*, given the two testimonies, is the following:

$$p(H|G,W) = p(H)\frac{\sum_{i \in \{\text{true,false}\}} p(G|S=i)p(W|S=i)}{p(G,W)}. \qquad (4.28)$$

From this simple scheme Pearl elaborated additional features, such as the prediction of a future event regarding Holmes' daughter, or the possibility that the alarm is triggered by an earthquake:

> Immediately after his conversation with Mrs. Gibbon, as Mr. Holmes is preparing to leave his office, he recalls that his daughter is scheduled to arrive home at any minute. If greeted by an alarm sound, she probably (P = 0.70) would phone him for instructions.

> [...]

> Mr. Holmes remembers reading in the instruction manual of his alarm system
> that the device is sensitive to earthquakes and can be accidentally triggered by
> one. He realizes that if an earthquake had occurred, it surely would be on the
> news.
>
> (Pearl, 1988, p.47,49)

The graph with the new additions is shown in Figure 4.12 on the right. As in most cases of real reasoning, there are now multiple possible causes: S can be true because of a burglary H, or because of an earthquake E. This alternative explanation of S can be reinforced by the evidence R, news reporting the earthquake. There is also an event D in the future, the call of Holmes' daughter, that can be predicted depending on the other evidences.

It can be said that Bayesian networks stand somehow in between the two main poles in AI: rationalism and empiricism. What is shared with rationalism is the application of abstract rules, in this case Bayes' theorem. However, the rules can be adapted within their parameters by learning, with Bayesian update of probabilities at every new evidence. In fact Geoffrey Hinton appreciated the work of Pearl, and derived a sort of neural version, described in §5.1.2. Conversely, Pearl stated that the idea of Bayesian networks came to him from an article by David Rumelhart, adding, "I wanted Bayesian networks to operate like the neurons of a human brain" (Pearl and Mackenzie, 2018, p.121).

Bayesian based AI has undergone several developments and refinements in the last thirty year (Bovens and Hartmann, 2003; Koller and Friedman, 2009; Korb and Nicholson, 2011). It is still a valid framework for knowledge engineering; however, it has never triggered a significant advancement in AI, comparable to that in other directions like artificial neural networks. This rather limited impact is acknowledged by Pearl himself, in his most recent book:

> [...] I became a convert to causality through my work in AI and particularly on
> Bayesian networks. These were the first tool that allowed computers to think
> in "shades of gray" – and for a time I believed they held the key to unlocking
> AI. Toward the end of the 1980s I became convinced that I was wrong, and this
> chapter tells of my journey from prophet to apostate.
>
> (Pearl and Mackenzie, 2018, p.23)

The idea that Bayes' paraphernalia can be successful in explaining the mind has not only never waned, but has found a new impulse today, thanks to modern statistical tools known as variational Bayes. We will tell more about the return of Bayes in §5.1.3.

4.4.2 Darwin

Like Thomas Bayes, Charles Darwin was not a philosopher, but neither was he an amateur scientist; he was one of the greatest naturalists ever. The vast body of studies and theoretical reflections by Darwin (1880, 1859, 1871, 1872), with *On the Origin of Species* at the forefront, established a new form of explanation for the history and diversity of life on Earth called *Darwinism*. Like Bayesianism, Darwinism is, in its own right, a philosophical vision, but unlike Bayesianism it is so broad and well known that an introduction here is useless (see Mayr, 1963; Dawkins, 1976; Grene, 1983; Brandon, 1990; Williams, 1992; Brandon, 1996).

Darwinism, well before being taken into consideration by AI, had had a strong impact on cognitive science with the approach known as *evolutionary psychology*. According to evolutionary psychologists, cognitive mechanisms are adaptations – outcomes of natural selection – that allowed our ancestors get around the world, survive and reproduce. For the leading evolutionary psychologists like Leda Cosmides and John Tooby (Barkow et al., 1995; Cosmides and Tooby, 1997; Tooby and Cosmides, 2000), and Steven Pinker (1997), the study of cognitive processes as natural adaptations provided a unifying foundation for explaining all sorts of human behaviors. Others, like Daniel Dennett (1995, 2017), have argued for a mixed contribution of genetic and cultural evolution in shaping the human mind. Radical positions such as those of Cosmides and Tooby have been met with strong skepticism for ignoring a wide variety of other explanations of human behavior, most of which are not reducible to genetic adaptation (Little, 1991; Tomasello, 1999; Panksepp and Panksepp, 2000; Downes, 2005). In its radical form, evolutionary psychology entails strong innateness, which lead Pinker to say that language is an 'instinct', thus engaging in a strenuous fight against empiricist AI (see §4.2.1). Moreover, philosophers of biology even if sympathetic with the general idea of evolutionary psychology, have raised severe criticisms for a loose, if not inappropriate, use of fundamental concepts of biological evolution, like fitness (Griffiths, 1996; Grantham and Nichols, 1999). Robert Richardson (2007) dismisses evolutionary psychology and its flawed methods as 'maladapted psychology'.

Darwinism found its way into AI at the end of the 1960s much more softly and quietly than did evolutionary psychology. Several groups around the world were captivated by the idea of imitating natural selection for solving computer problems, each with independent and slightly different directions. The group lead by Ingo Rechenberg (1973) at the Technical University of Berlin proposed the *evolutionary strategies*, Lawrence Fogel et al. (1966) at the University of California at Los Angeles proposed *evolutionary programming*, while John Holland (1962) at the University of Michigan in Ann Arbor worked out the field of *genetic algorithms*. These approaches were ignored by mainstream AI and have since been almost forgotten, with the notable exception of genetic algorithms; hence, we will only describe the latter. The inception of genetic algorithms was slow and quiet, and for a long while it progressed almost entirely thanks to the work of the group at Ann Arbor lead by Holland (1975). By the end of the 1980s, genetic algorithms found a moderate worldwide interest, and a small research community started to grow (Jong, 2005).

Informally, genetic algorithms can be described as follows: Given a problem for which there can be many possible solutions but most of them are very poor, each solution is represented by an individual in a population, and the population is evolved for generations, leaving only the best solutions to survive.

The road from this simple description to a working algorithm is not easy at all, and has required the implementation of several basic concepts of Darwinism as computing entities; a list of the main concepts necessary is in Table 4.1. For the sake of clarity, we will now discuss all these concepts in a real problem for which a genetic algorithm has been used: the design of a building facade (Wright and Mourshed, 2009; Gagne and Andersen, 2012). The problem is to design facades on all sides of a building, successful in creating a daylighting scheme on the interior, with a pleasant illuminance level and avoiding annoying glare. The objective is difficult, taking into account that the solution should be optimal all the year, and for all daytime hours.

An individual organism in nature corresponds now to a solution: a full specification of all facades for the building under design. The set of solutions currently under analysis in the algorithm corresponds, in biological terms, to a population. In nature each individual organism has a chromosome, the design of the corresponding mathematical structure is a central issue. This structure should somehow code all features that characterize one solution to the problem at hand. A single feature corresponds to one biological gene. Typical genes/features in our case might be the window-to-wall ratio; the number of windows; their locations; the use of overhang or fins; the glass transmissivity and reflection, and so on.

The coding initially proposed by Holland (1975) used sequences of bits with fixed length, and a specific coding language; later on, more flexible strategies have been introduced (Michalewicz, 1994; Schmitt, 2001), allowing genes to be single bits, numbers (Eshelman and Schaffer, 1993), or even generic programming objects. In the case of building facades a feature like 'overhang' can be coded by a single on/off bit, while window-to-wall ratio is better coded by a real number.

Table 4.1: List of concepts in Darwinism and corresponding components or processes in genetic algorithms.

Darwinism	genetic algorithm
organism	single solution
population	set of solutions
chromosome	coding vector
gene	coding vector element
allele	coding element value
phenotype	set of parameters decoded from the coding vector
fitness	function from set of parameters to real
reproduction	crossover: random combination of two coding vectors
mutation	random perturbation of a coding vector
selection	subset of solutions ranked by fitness

From the chromosome/coding vector, the genetic algorithm should be able to construct the phenotype, in this case a full facade, typically within a 3D computer graphics software. The most crucial issue in the algorithm is the fitness. It is a function that should take as input a phenotype, and return a real value that measure how good the solution is. In our case, the fitness is evaluated by simulating the exposure of the building to a variety of daylight conditions over a year and over different times of the day, and measuring several aspects of the internal illumination, taking into account also other aspects such as enegy saving.

With the appropriate coding for chromosomes, and an adequate fitness function, the Darwinian evolution can take place with the cycle illustrated in Figure 4.13. There is an initial population with chromosomes generated randomly, taking into account the range of possible values for each gene/feature. Each individual in the population is evaluated for fitness, and the most fit individuals are selected. The individuals going into the selection pool will create offspring for the next generation, by mating in couples. Unlike in nature, individuals living inside genetic algorithms have no gender, so each can mate every other. The imitation of nature is in constructing the new chromosome for the offspring as a recombination of those of the parents'.

There are several possible recombination schemes (Sivanandam and Deepa, 2008); one of the most popular and simple is the single-point crossover (Deb and Agrawal, 1994), illustrated in Figure 4.14. A cross site within the chromosome is chosen at random, and the chromosomes of the parents are cut along this cross. The fragments of chromosomes on the left and on the right of the cross site are swapped to produce two children.

Genetic algorithms are an effective alternative to both rationalist and empiricist view in AI. There are no logical rules like in rationalist models, nor there is any form of learning from experience, as in empiricist models. This last point is an advantage for applications where a history of past experiences is scarce or completely lacking. This is exactly the case in our example on facade design.

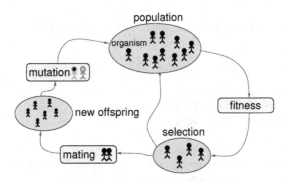

Figure 4.13: General scheme of the genetic algorithm cycle.

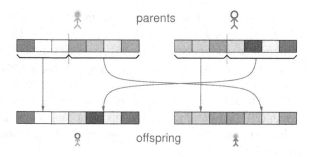

Figure 4.14: The single-point crossover technique used for mating parents.

There is an enormous number of facades in the world, but there are no datasets of parameters of the facades and related daylighting performances.

Although genetic algorithms and evolutionary psychology are offspring of the same parent, Darwinism, they have never crossed paths. The genetic algorithm community has had no relation at all with cognitive science, for obvious reasons. The form of artificial intelligence provided by genetic algorithms do not mimic any individual cognitive intelligence, but some kind of 'intelligence' indirectly emerging from the action of natural selection on a population of individuals. On the other hand genetic algorithms have intertwined with domains where interactions within a population are crucial, such as economy (Arrow, 2005) and Artificial Life (Mitchell and Forrest, 1994).

The fair success of genetic algorithms ignited, in the 1990s, a wave of research on what biology could offer, beyond Darwinism, that could be borrowed for algorithms. One of the first such attempts was *particle swarm optimization* (Eberhart and Kennedy, 1995), inspired by the efficient swarm behavior of fish or birds in nature. Soon after, Dorigo et al. (1996) proposed the *Ant Colonies*, based on the metaphor of ants that are able to find the best path to travel by following pheromone previously deposited by other ants. It is now hardly possible to find a social animal that has not been a source of inspiration for algorithms, including fireflies, cuckoos, flower pollinating insects, bats (Yang and He, 2015). Despite the natural curiosity for their natural inspiration, these algorithms add little to the strengths of genetic algorithms, which remains the most reliable method of the biological family within AI.

4.4.3 Heidegger

Unlike Bayes and Darwin, Martin Heidegger was an all-round philosopher. His ideas, loosely ascribed to phenomenology, hermeneutics and existentialism, have been a major contribution to contemporary Continental philosophy. The influence of Heidegger in AI is, surprisingly, an American story. Heidegger certainly had no interest in connections with the American intellectual discourse; on the contrary, he perceived the United States as a major threat to all things German, a *Katastrophenhaft* (cataclysm), in the words of Ceaser (1997). This sentiment was not fully reciprocated, and indeed for several American students in philosophy Heidegger was an European attraction. A few of these students even traveled across the Atlantic to Freiburg to attend his lectures. It is due to these students that Heidegger started to have reception in America, as per the in depth analysis by Martin Woessner (2011). Not all those students went back across the Atlantic infused with enthusiasm and admiration, like one of the first, Marjorie Grene (1948), bitterly critical of Heidegger.

Other American students were fascinated by Heidegger, among them John Wild, an ardent anti-communist advocating existentialism as the best weapon to fight the doctrines of socialism (Wild, 1955). Here lies the connection with AI. Hubert Dreyfus, a brilliant student of Wild, was imbued with a firm admiration for Heidegger. So Dreyfus made his pilgrimage to Freiburg between 1953 and 1954, and when back in Harvard, helped Wild in the translation of the first part of *Sein und Zeit* (Heidegger, 1927). Dreyfus's reshaping of the Heideggerian prophecy for the American audiences, later reconstructed by Woessner, is remarkable. Grene (1976), in dividing the attitude in American Heidegger scholarship between 'the simple adorers of Heidegger' and those who applied Heidegger 'towards the development of their own reflections', had already listed Dreyfus among the latter. John Searle, with his typical humor, stated that his friend Dreyfus worked not so much on Heidegger, but 'Dreydegger'. An obvious example of Dreyfus's narrow perspective on Heidegger is his insistence on *Sein und Zeit* only, even in the first division of this work (Dreyfus, 1990). By focusing almost exclusively on this part of the whole philosophical production of Heidegger, Dreyfus can offer only what he is interested in: a critique of traditional ontology and epistemology, from Cartesian rationality up to logical positivism. And Dreyfus identified in AI, the latest and most poisonous outcome of this tradition, against which Heidegger would be the best antidote.

> Thus both philosophy and technology finally posit what Plato sought: a world in which the possibility of clarity, certainty, and control is guaranteed; a world of data structures, decision theory, and automation. No sooner had this certainty finally been made fully explicit, however, than philosophers began to call it into question. Continental phenomenologists recognized it as the outcome of the philosophical tradition and tried to show its limitations. [...] Heidegger calls it *rechnende Denken*, "calculating thought", and views it as the goal of philosophy, inevitably culminating in technology. Thus, for Heidegger, technology, with its insistence on the "thoroughgoing calculability of objects", is the inevitable culmination of metaphysics, the exclusive concern with beings (objects) and the concomitant exclusion of Being (very roughly our sense of the human situation which determines what is to count as an object).

> (Dreyfus, 1972, p.124)

The acrimony shown by Dreyfus towards AI was already expressed in an early report for the RAND Corporation where the scientific status of AI was compared to alchemy (Dreyfus, 1965), and the dismissal of AI on Heideggerian basis was restated years later (Dreyfus, 1992).

If the result of the American reception of Heidegger had been simply of rejection of AI, it would not be useful to talk about it here. But the capacity of intellectual assimilation on the part of America produces unexpected effects, even the interweaving of the austere German philosopher with well known American entrepreneurship (Collins and Moore, 1964; Weiss, 1969; Jones and Spicer, 2009; see also Okrent, 2003). The key protagonist, Fernando Flores, is actually not North American, he is a successful Chilean entrepreneur, former finance minister in the government of Salvador Allende. He was imprisoned after the military coup of Pinochet and released in 1976, thanks to the good offices of the San Francisco chapter of Amnesty International, which brought him to Berkeley, where he meet Heidegger under the interpretation of Dreyfus. Under Dreyfus's guidance, Flores obtained his PhD on the strange subject of Heidegger and office work. The theme was further pushed as a sort of American self-help philosophy in a book authored by Flores together with Dreyfus and Spinosa, *Disclosing New Worlds: Entrepreneurship, Democratic Action, and the Cultivation of Solidarity*, that promised to teach the reader 'to appreciate and engage in the ontological skill of disclosing new ways of being' (Spinosa et al., 1997, p.I) Flores set up several companies selling 'Heideggerian' communications software, conducting seminars and giving managers instructions influenced by Heidegger; one of these companies was named "Hermenet". Flores describes the application of Heidegger in another of his companies, BDA, as follows:

> With Dreyfus's Heidegger as a cornerstone of our thinking, I and my colleagues at Business Design Associates (BDA) have been developing, and applying with our clients, an alternative description of business that both challenges the interpretations held by most business consultants and people trained in U.S. business schools and leads to significant increases in competitiveness.
>
> (Flores, 2000, p.271)

The "applied" Heidegger movement has been described well by Anthony Gottlieb in The New York Times (Jan 7, 1990) in the article *Heidegger for Fun and Profit*. Flores had been the key player in the infiltration of Heidegger into AI itself. He succeeded in convincing one of the founders of AI, Terry Winograd (see §3.2.3), to entirely abandon his previous ideas, turning to hermeneutics, phenomenology, and autopoiesis too, after the readings of Flores' friend and compatriot, Maturana. Winograd himself could hardly have imagined his volte-face; in 1968 he wrote a memo in response to Dreyfus's bitter attacks on AI titled *The Artificial Intelligence of Hubert L. Dreyfus: A Budget of Fallacies*. Then, convinced by Flores, Winograd embraced the arguments of Dreyfus, once thought of as fallacies, in debunking AI. However, he still believed that AI, if able to turn into a Heideggerian vision, may survive. The book coauthored by Winograd and Flores mixes a fair amount of criticism with the tiny hope of a possible future of a 'Heideggerian AI' Winograd and Flores (1987). But before this promised future AI, the book was a very brutal shock for the AI of that period: Winograd and Flores rejected the computational approach in its entirety, and functionalism was declared intellectually bankrupt. Margaret Boden (2008, p.857) narrates the evolution of Winograd's position in a section titled *The Unkindest Cut of All*: "In the eyes of the GOFAI community, this was betrayal. Winograd had started as a heretic and rapidly became a high priest of the orthodoxy. Now, he was an apostate."

As a matter of fact Winograd, after his conversion to Heideggerian wisdom, did not produce anything even remotely comparable to his previous research. However, the pragmatic attitude of Winograd and Flores mitigated the view of AI as the exemplar of the abominable dehumanization of technology, congruent with Heidegger's philosophy and espoused by Dreyfus. Winograd worked on introducing Heidegger into the MIT AI Laboratory and, in 1986, Dreyfus himself was invited for a seminar called "Why AI Researchers should study *Being and Time*". The practical messages distilled from the Dreydeggarian rhetoric for AI researchers were, not to rely entirely on symbolic representations, to move from the detached perspective to a situated one, and to forget model-based planning in favor of action planning emerging from concrete interactions with the environment.

Two young researchers at the MIT AI Lab were particularly receptive to this kind of message: Philip Agre and David Chapman. They attacked, in particular, the standard model-based planning of AI, claiming Heidegger as inspiration:

> We want to understand the emergence of abstract reasoning from concrete activity. We believe that abstract reasoning is not innate, but derived from concrete activity [...] Heidegger argues that the phenomenologically primordial way of being is involvement in a concrete activity. Everyday activity consists in the use of equipment for a specific reason.
>
> (Chapman and Agre, 1986, p.412)

Agre and Chapman (1987) succeeded in developing a program implementing their theoretical stance, called *Pengi*. It is a virtual agent playing a computer game called *Pengo*, where a penguin navigates a maze made of ice blocks, chased by killer bees that can be killed by sliding an ice block into it. Instead of relying on a rule-based plan, "Pengi lives in the present, continually acting on its immediate circumstances." (p.269). Agre and Chapman had to make use of representations, but to mark the difference from the classical symbolic representations of AI, they called them *deictic representations*.

Unfortunately, deictic representations found no future beyond Pengi. The only significant advancement in AI, loosely related to Heidegger/Dreydegger wisdom, is due to Rodney Brooks. Australian, Brooks joined MIT in 1984 where he lead the mobile robot group. A few years later Brooks (1990b) published a highly provocative paper with the catchy title, *Elephants Don't Play Chess*. The paper raises a strong challenge to the symbol system hypothesis, and to a number of additional aspects which it implies, like the notion of a central intelligence, and the so-called *frame problem* where, in the system, it is impossible to assume anything that is not implicitly stated. Brooks did not rest in dismantling the artificial intelligence, he elaborated an engineered alternative called *subsumption architecture*. It is based on synchronous modules named *layers* (Brooks, 1986) that instantiate simple computational machines connected to perceptual sensors and control motor actuators. There is no central locus of control; layers are made by finite state machines that communicate with each other by passing messages that cause the change of their internal states. New layers can be dynamically added to achieve new behaviors, and higher level layers can 'subsume' the roles of lower levels which, however, continue to function. Brooks (1990a) designed a specific programming language, called *behavior language* for the subsumption architecture. By 1990, a number of increasingly complex robots based on the subsumption architecture were built at MIT Lab. The first one, Allen, was just able to wander about avoiding obstacles; in addition to this, Herbert could search for a soda can and eventually grasp it; Seymour was capable of visually tracing moving objects, while moving itself.

Despite the similarities between Dreyfus's criticism of symbolic AI and that of Brooks, the latter preferred to stand apart from the Heideggerian trend:

> In some circles much credence is given to Heidegger as one who understood the dynamics of existence. Our approach has certain similarities to work inspired by this German philosopher [...] but our work was not so inspired. It is based purely on engineering considerations.
>
> (Brooks, 1991, p.155)

By the end of the last century, the new robotics inspired by the subsumption architecture spread around the world, with many examples of robots reaching capabilities close of those of simple animals like insects, previously unimaginable. According to Boden (2008, p.1031) some jokers remarked that 'AI' now stood for 'Artificial Insects'. Brooks (1997, 1999) planned to move much further, with a long term robotics project aimed at reproducing aspects of of human cognition. The first prototype, *Cog*, has a fairly humanoid-like upper-torso structure, and one of its most peculiar feature is joint attention, by mutual gaze, gaze following, and pointing. As argued by Dennett (1997), Cog has the right basis for showing a sort of artificial theory of mind, one of the most advanced human capabilities.

Cog paved the road for the current cognitive robotics, where several paradigms like embodiment, neural networks, and social cognition meet together, but where Heideggerian AI is almost dissolved, even if still remarked as a landmark in the history of robotics (Bhaumik, 2018). Still, Dreyfus (2007) insists that

Heideggerian AI, like that implemented in Brook's robots, is not yet giving satisfactory results because ... it is not Heideggerian enough. Others, like Herrera and Sanz (2016), have argued that Heideggerian AI, as conceived by Dreyfus, clashes with Heidegger's philosophy, and using Heidegger for the advancement of AI is a contradiction in its own terms.

Chapter 5

An Unexpected Renaissance Age

AI is currently in its most exciting period ever. Several authors, like Tan and Lim (2018), used the term *AI Renaissance* to delineate this fortunate trend. The parallel with the flowering of culture in Italy that begun in the 14th century is somehow appropriate, because it characterizes a flourishing recovery following a dark period. In fact, as we will describe in the first section of this chapter, AI has a troublesome history, plagued by periodic negative phases known as *AI winters*. Even the term *AI Renaissance* was not invented recently; this definition has already been used in the past when there were signs of coming out of one of the winter seasons (O'Leary, 1997, 2001). The current Renaissance age is almost entirely due to a family of algorithms called *Deep Learning* (DL) which will be introduced in section §5.1. There is no comparison between the current brilliant moment in AI and all its previous Renaissance periods; a detailed discussion of this success will be given in section §5.2. We will address in section §5.1.4, a surprising aspect of this success: not only was it sudden and totally unexpected, but there is still no satisfactory explanation of why it happened at all. DL has progressively dominated every area of AI, but there is one area in particular where it is surprisingly predominant: artificial vision. The last section of this chapter is focused on this case, constrained with natural language processing, where DL emerged later and at a slower pace.

5.1 The Warmth of *Deep Learning*

To some extent, the pattern of AI development is not unusual: it is common for emerging technologies to follow a trend that is anything but linear. A popular tool in business management is Gartner's *Hype Cycle* (Fenn, 2007), a sort of universal curve that would describe the evolution in time of the visibility of a new technology. The plot has four stages. The first, *Technology Trigger*, is when the technology first shows up. Then, the buzz about it raises the curve up to the *Peak of Inflated Expectations*, soon followed by the *Trough of Disillusionment*, when people realize the new technology really does not deliver a revolution. That low point is followed by the *Slope of Enlightenment*, associated with the beginning of real adoption growth of the technology, in a limited number of applications. This rising slope ends in the flat, final stage of the *Plateau of Productivity*.

5.1.1 Cold Seasons

AI, however, has followed a rather different pattern, and instead of a single negative period, such as Gartner's *Trough of Disillusionment*, shows periodic negative peaks, reminiscent of the cold seasons. And while the *Peak of Inflated Expectations* in Gartner's plot is mainly driven by the media and by technology users, high expectations in AI are often promised by its leading researchers.

These were the words of Nobel laureate and AI pioneer Simon, in the early days of AI:

> The simplest way I can summarize the situation is to say that there are now in the world machines that think, that learn and that create. Moreover, their ability to do these things is going to increase rapidly until – in a visible future – the range of problems they can handle will be coextensive with the range to which the human mind has been applied.

> (Simon and Newell, 1958, p.8)

Similar remarks were echoed by other AI leading researchers, as evidenced by the few below:

> Programming computers to play games, to write poetry and solve high school problems is but one stage in the development of an understanding of the methods which must be employed for the machine simulation of intellectual behavior. [. . .] it seems certain that the time is not far distant when most of the more humdrum mental tasks, which now take so much human time, will be done by machine.

> (Samuel, 1962, p.11)

> Computers's eyes once could sense only a hole in a card. Now they recognize shapes on simple backgrounds. Soon they will rival man's analysis of his environment. Computer programs once merely added columns of figures. Now they play games well, understand simple conversations, weigh many factors in decisions.

> (Minsky, 1968, p.3)

Such confidence in the future of AI encouraged substantial funding, especially in USA and in the UK, until eventually, it was clear that AI had been oversold. A first severe AI winter arrived in 1973 to freeze the excitement. The trigger was the resounding rise of AI at the University of Edinburgh, thanks to Donald Michie. During the Second World War, he was one of the cryptographers at Bletchley, together with Alan Turing, and had come to believe in AI to the point "to make machine intelligence my life" (Michie, 2002). With his energy and commitment Michie established the Department of Machine Intelligence and Perception at Edinburgh in 1966, and soon become a leader of AI in the UK (pp.348–350 Boden, 2008). The University of Edinburgh became a sort of European cradle of AI, attracting students like Goffrey Hinton, which will be addressed in detail later. Unfortunately, Michie was no less over-optimistic than the AI leaders cited above, and his often abrasive personality provoked discord within the scientific community at Edinburgh and in the UK in general. In the face of this situation, in 1971, the UK Science Research Council, the main founder of AI research in UK, commissioned the mathematician Sir James Lighthill to investigate AI in general and the Edinburgh group in particular. The report was extremely negative, as evident from this passage:

> Most workers in AI research and in related fields confess to a pronounced feeling of disappointment in what has been achieved in the past twenty-five years. Workers entered the field around 1950, and even around 1960, with high hopes that are very far from having been realised in 1972. In no part of the field have the discoveries made so far produced the major impact that was then promised.
>
> (Lighthill, 1973, p.8)

The result was devastating for AI, soon UK funding all but vanished, and gradually, in the United States too, bold ARPA initiatives in AI were scaled back. Unlike the calendar season, this AI winter lasted for over ten years.

By the early 1980s, the ice had thawed. The spring for AI came in the form of a challenge from Japan in 1981, announcing a ten-year national plan for developing *Fifth Generation* computers, designed to be suitable for AI applications. This announcement had a huge effect worldwide, with Western governments and venture capitalists fearing a future Japanese dominance in computing. Feigenbaum and McCorduck (1983) wrote a book with a significant title: *The Fifth Generation: Artificial Intelligence and Japan's Computer Challenge to the World*, arguing that hugely increased funding for AI was needed if USA were to compete with Japan. In 1983, US DARPA funding for AI ramped up sharply, and in 1984, Europe steered its ESPRIT programme toward AI for over 200 million dollars per year.

One again, it was hard for AI protagonists to refrain from echoing the enthusiastic and over-optimistic claims of the early days: machines would soon reach, and possibly surpass, human intelligence. Here is an example:

> Fully intelligent machines will result when the metaphorical golden spike is driven uniting the two efforts. A reasoning program backed by a robotics world model will be able to visualize the steps in its plan, to distinguish reasonable situations from absurd ones, and to intuit some solutions by observing them happen in its model, just as humans do. [...] I expect to see this union in about forty years.
>
> (Moravec, 1988, p.3)

Another bubble was forming; in the late 1980s, DARPA has spent 2 billion dollars in AI with very limited results, far from expectations, and a second AI winter arrived. Funding for AI dried up again, all over the world.

One might wonder why distinguished AI scientists have repeatedly made such unrealistic announcements, and an interesting answer is offered by Jerry Kaplan, himself one of these scientists:

> I'm an optimist. Not by nature, but by U.S. government design. After Russia humiliated the United States with the 1957 launch of Sputnik, the first space satellite, the government decided that science education should be a national priority. [...] Young boys like me (but tragically, not many girls) were fed a steady diet of utopian imagery extolling technological innovation as the path to eternal peace and prosperity, not to mention a way to beat them clever Russkies.
>
> (Kaplan, 2015, p.3)

At the turn of the millennium the over-optimism "by U.S. government design" chronicled by Kaplan should only be a faint memory, but in the meantime the waves of hype followed by disappointment have eroded the reputation of AI. A stable cold climate developed around AI, with big drop off in investment. Attendance at AAAI (Association for the Advancement of Artificial Intelligence), the biggest worldwide AI conference is a useful indicator of the liveliness of the field. It dropped from 5000 in 1988 to 3000 in 1990, 2000 in 1991, and stabilized to slightly under 1000 at the turn of the millennium. There were, in fact, several instances of slow but steady progress in various AI domains, but the field, as a whole, was so discredited that researchers preferred to avoid the "Artificial Intelligence" label altogether. The AI expert, James Manyika, recalls that period with these words:

> [...] they were actually working on AI, but not many people called it that in those days because AI had a negative connotation at the time [...] So, they called their work everything but AI – it was machine perception, machine learning, it was robotics or just plain neural networks; but no-one in those days was comfortable calling their work AI.
>
> (Ford, 2018, p.294)

Even in a field like big data mining that, today, is considered entirely part of AI, there are striking recent examples of the reluctance to be labeled as AI. In one of the most authoritative reading for students and practitioners in big data mining (Rajaraman and Ullman, 2011) there is no a single occurrence of 'artificial intelligence'; similarly in (Han et al., 2012) there is just one occurrence, but only in the introduction. It can be argued that in these cold winter periods, the negative meaning of 'artificial', denoted in the §1.3, has overwhelmingly resurfaced.

5.1.2 From Shallow to Deep

Now the reputation of AI has reached its zenith, and – continuing James Manyika's quote of the last section – "[...] Now we have the opposite problem, everyone wants to call everything AI" (Ford, 2018, p.294). Today, even sex toy companies find that adding the 'AI' label makes their flagship products more marketable, like the 'Autoblow-AI' device, manufactured by Very Intelligent Ecommerce Inc.

The reversal of AI's reputation has happened due to DL. Remarkably, DL revolutionized AI without being a revolution. It is a smooth advancement of ANN, of the same models proposed by the PDP project described in §4.1. Even the leading researchers of DL are the same as those of the old PDP project, primarily, Goffrey Hinton, often known as the Godfather of DL. The term, 'learning', in 'deep learning' unequivocally flags the empiricist philosophy behind DL, in direct continuity with ANN: models derive all their capabilities and their intelligence by learning from experience.

The novelty is in the term 'deep'. Probably the most prominent representatives of deep learning would not dislike the metaphorical usage of the adjective 'deep' as intellectually profound, capable of entering far into a matter. But its meaning is actually technical and rather trivial, it refers to the number of hidden layers in a feed-forward ANN. Neural models can learn increasingly complex functions by augmenting the number of units; this way, however, the number of parameters to optimize increases as well, and learning becomes more difficult. Units can be added using two different options: by increasing the number of units in the existing layers, or by adding new layers.

In the 1990s there was a widespread belief that increasing the number of units by adding layers was much less efficient than increasing the width of a single hidden layer. This fact was based on several empirical observations, like those of de Villers and Barnard (1992), whose conclusion is as follows:

> We have found no difference in the optimal performance of three- and four-layered networks [...] four layer networks are more prone to the local minima problem during training [...] The above points lead us to conclude that there seems to be no reason to use four layer networks in preference to three layers nets in all but the most esoteric applications.

The recommendation to use no more than one hidden layer has gradually become an undisputed dogma, and all 'but the most esoteric' feed-forward ANN were made of three layers: input, hidden and output, models that now are called 'shallow'. The sense of 'deep' in DL is the revision of this dogma. The point is that the observed difficulties in training four layer networks, i.e., with two hidden layers, were not due to an intrinsic advantage of having wide single hidden units with respect to many smaller layers. It was the standard backpropagation learning algorithm that worked quite well with models with one hidden layer and lost efficiency with more hidden layers. Intuitively, the main reason lies in the computation of the gradient of weights with respect to the errors in Equation (4.5). This computation requires using the chain rule and the need to multiply each layer's weight and gradients together across all the layers. With just one hidden layer this computation is not critical, but with many layers, there is the problem known as 'vanishing gradients'. If most of the weights across many layers are less than 1, since they are multiplied many times, eventually the gradient just vanishes into the smallest machine number, and the training stops.

The first 'deep' way out of the vanishing gradient problem was proposed by Hinton and Salakhutdinov (2006), who were able to train models with four and five hidden layers. This result was achieved by dusting off one of Hinton's beloved ideas: Restricted Boltzmann Machines (RBM), described in §4.1.3. While backpropagation works in supervised mode, RBM understand the data unsupervisedly. Hinton's clever trick was to take two adjacent layers in a feedforward network, and train them as Restricted Boltzmann Machines. The procedure starts with the input and the first hidden layer, so that it is possible to use the inputs of the dataset to train the unsupervised Boltzmann Machine model. Then, this model is used to generate a new dataset, just by processing all the inputs. This new set is used to train the next couple of layers. This procedure is a sort of pre-training that gives an initial shape to all the connections in the network, to be further refined by ordinary backpropagation using both, the inputs and the known outputs of the dataset. Figure 5.1 illustrates the method applied to the MNIST dataset of handwritten digits, one of the benchmarks on which the model has been tested. The full model is organized as 'autoencoder', a neural networks whose task is to reproduce its own input on output. Mathematically, the autoencoder can be written as the combination of two functions:

$$g_\Phi \ : \ \mathscr{X} \subseteq \mathbb{R}^M \to \mathscr{Z} \subseteq \mathbb{R}^N, \tag{5.1}$$

$$f_\Theta \ : \ \mathscr{Z} \subseteq \mathbb{R}^N \to \mathscr{X} \subseteq \mathbb{R}^M. \tag{5.2}$$

The first is called *encoder*, defined by a set of parameters Φ (all the weights of the neural connections), and it computes the inner representation $\vec{z} \in \mathscr{Z}$ of the input $\vec{x} \in \mathscr{X}$. The second function is the *decoder*, defined by the parameters Θ, and it aims to reconstruct the original data $\vec{x}$ from the inner representation $\vec{z}$. Therefore, when the autoencoder is well trained, one should expect the output to

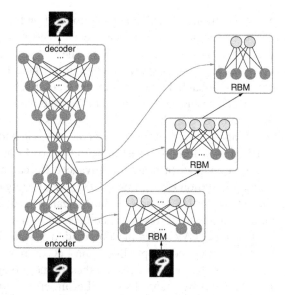

Figure 5.1: Sketch of how Hinton and Salakhutdinov (2006) train models with many layers. The left graph is the full model, organized as autoencoder, with the encoder made of a stack of layers with progressively fewer units, and the decoder with inverse sized layers. The graphs on the right are Restricted Boltzmann Machines (RBM) that include just two layers of the full model. The RBM with the first two layers is trained using as input the same images of the full model. The second RBM uses as input the output of the first RBM; once trained, the same, in turn, trains the other RBM.

be a close approximation of the input:

$$\vec{x}' = f_\Theta\left(g_\Phi(\vec{x})\right) \approx \vec{x} \tag{5.3}$$

This might sound like a stupid target, but the advantage comes from the progressive decrease in the size of the layers. Typically $M \gg N$, therefore the inner layer is a vector condensing in a few numbers, all the most previous features extracted by high dimensional inputs like images. In some of the experiments of Hinton & Salakhutdinov the input and output are images, 28×28, therefore $M = 784$, while the inner layer is as small as just $N = 6$. If the autoencoder, with many stacked layers, had to be trained with backpropagation in order to fix the sets of parameters Φ and Θ, it would have been a hopeless task.

This first neural model with many hidden layers launched the use of the attribute 'deep', in this case 'deep autoencoder', while in the simultaneous work by Hinton et al. (2006), the neural model was called *Deep Belief Net* (DBN). In this case the models are the sort of mixture between neural networks and Bayesian networks, described in §4.4.1, and the strategy to train such models with many layers is similar, learning the weight matrices one layer at a time, using the RBM learning algorithm. It was soon recognized that the layer-wise pre-training was an elegant way out of the problems afflicting backpropagation for any neural network with many layers (Bengio et al., 2007) and even RBM models themselves became 'deep' (Salakhutdinov and Hinton, 2009). The first practical application of unsupervised pre-training was in speech recognition, where DBNs achieved record-breaking results (Mohamed et al., 2011) and were quickly developed by many of the major speech groups and deployed in Android phones.

But the major outcome of deep belief pre-training was in promoting the new Renaissance in ANN, and gradually in AI as a whole. Hinton had found one of the rare research institutes willing to invest in neural networks at the beginning of the millennium, the Canadian Institute for Advanced Research (CIFAR). He had already left Carnegie Mellon University in 1987, for a variety of reasons. He was particularly upset by the US politics in the Reagan period and, not liking having to take military funds to do research, he decided to live in Canada. By early 2004, he was leading CIFAR's Neural Computation & Adaptive Perception program (NCAP), which attracted people like Yann LeCun, postdoc of Hinton, and Yoshua Bengio. The collaboration between these three protagonists on the rise of deep learning is told by themselves in (LeCun et al., 2015). CIFAR's investments in the dark period of artificial neural networks were far-sighted, and in 2017 raised up to half a million dollars a year.

5.1.3   Good Old Tools

Deep belief pre-training had the great merit of acting as a catalyst of ANN Renaissance, but from a technical point of view it was not so essential for DL. It turned out that good old backpropagation (see §4.1) can easily train deep feedforward networks, with just a few improvements, without the complex procedure of pre-training. The Renaissance atmosphere is also reflected at the linguistic level; today, as we prefer to talk about DL instead of ANN, the term 'backpropagation' has fallen into disuse, replaced by others. One of the most popular is *Stochastic Gradient Descent* (SGD) (Bottou and LeCun, 2004). Its basic formulation is the following:

$$\vec{w}_{t+1} = \vec{w}_t - \eta \nabla_w \frac{1}{M} \sum_i^M \mathscr{L}(\vec{x}_i, \vec{w}_t) \qquad (5.4)$$

and one can instantly see the similarity with the standard backpropagation equation (4.5). Instead of computing the gradients over a single sample t, in Equation (5.4) a stochastic estimation is made over a random subset of size M of the entire dataset, and at each iteration step t a different subset, with the same size, is sampled.

The name SGD gives credit to a mathematical context rather different from the one to which the backpropagation belongs (see §4.1), that of stochastic approximation, established by Robbins and Monro (1951). The idea is to solve the equation $f(\vec{w}) = \vec{a}$ for a vector $\vec{w}$, in the case when the function f is not observable, using samples of an auxiliary random function $g(\vec{w})$ such that $E[g(\vec{w})] = f(\vec{w})$. The solution is obtained by the following iterative equation:

$$\vec{w}_{t+1} = \vec{w}_t - \frac{\alpha}{t}\left(g(\vec{x}_t) - \vec{a}\right). \tag{5.5}$$

Stochastic approximation was mostly developed in engineering domains, and has turned into an extensive mathematical discipline (Kushner and Clark, 1978; Benveniste et al., 1990). Many minor variants on SGD have been introduced during the ANN Renaissance; for example, by adapting the learning rate η during learning (Duchi et al., 2011), or mixing the update given by Equation (5.4) at step $t + 1$, with the upgrade computed at the previous step (Kingma and Ba, 2014).

Hinton has never concealed his definite preference for unsupervised learning – as in RBM – over backpropagation. Nevertheless, just as he contributed significantly to the early development of backpropagation, he added an important innovation to SGD too Hinton et al. (2012). His technique, called *dropout*, randomly switches off a fraction of the neurons during training. This trick prevents a sort of unwanted co-adaptation of feature detectors, driven by minimizing the error on a set of samples. With dropout, neurons learn to detect a feature that is generally helpful in producing the correct answer for a large variety of possible inputs. Hinton suggests a curious analogy between dropout in DL models and the role of sex in evolution:

> One possible interpretation of the theory of mixability [...] is that sex breaks up sets of co-adapted genes and this means that achieving a function by using a large set of co-adapted genes is not nearly as robust as achieving the same function, perhaps less than optimally, in multiple alternative ways, each of which only uses a small number of co-adapted genes. This allows evolution to avoid dead-ends in which improvements in fitness require co- ordinated changes to a large number of co-adapted genes.
>
> (Hinton et al., 2012, p.6)

Regardless of the accuracy of this analogy, dropout turned out to be extremely attractive for training DL models.

Backpropagation is in good company among old methods revived by DL, as also those collected from philosophical inspirations different from the empiricist agenda of artificial neural networks. An interesting influence comes from the Bayesian perspective in AI, introduced in §4.4.1. Bayesian networks have been a precursor of sorts of DL, with the *Deep Belief Net* seen before (Hinton et al., 2006). The legacy of the Bayesian tradition exerted considerable influence when mixed with the variational methods. Originating in the 18th century with the work of Euler and Lagrange, the calculus of variations is concerned with finding how the value of a functional (a function of function) changes in response to variations of the input function. Variational Bayes method applies variational calculus to the Bayesian inference of conditional probabilities, expressed in Equation (4.27). It is a fertile field in statistical physics (Parisi, 1988; Blanchard and

Brüning, 1992) and signal processing (Šmídl and Quinn, 2005), but has found its way into DL too. In two concurrent and unrelated developments, Kingma and Welling (2014); Rezende et al. (2014), mixed the idea of autoencoders with that of variational inference; this new approach quickly became popular under the term *variational autoencoder*. Under the variational inference framework, the output $\vec{x} \in \mathscr{X}$ of the encoder – see Equation (5.2) – is a random variable, with probability distribution $p(\vec{x})$. It is not too difficult to shift from the deterministic neural network decoder f_Θ to a probabilistic function in the following way:

$$p_\Theta(\vec{x}|\vec{z}) = \mathbb{E}_{\vec{z} \sim \mathscr{N}(\boldsymbol{\mu}, \boldsymbol{\sigma})}\left[f_\Theta(\vec{z})\right] \tag{5.6}$$

where $\mathscr{N}(\boldsymbol{\mu}, \boldsymbol{\sigma})$ is the Gaussian distribution with mean $\boldsymbol{\mu}$ and standard deviation $\boldsymbol{\sigma}$. The desired approximation of $p(\vec{x})$ is, therefore, the following:

$$p_\Theta(\vec{x}) = \int p_\Theta(\vec{x}, \vec{z})\, d\vec{z} = \int p_\Theta(\vec{x}|\vec{z})\, p(\vec{z})\, d\vec{z}. \tag{5.7}$$

In practice, the variational autoencoder is much like the deterministic autoencoder, except that the inner layer has two values for each neuron, one for the mean, the other for the standard deviation of a Gaussian distribution. There are as many Gaussian distributions as the number of neurons, and the values fed to the decoder are just sampled from all these distributions, as shown in Figure 5.2.

We saw in §4.4.1 that the so-called 'Bayesian epistemology' – the idea that our reasoning is probabilistic in accordance to Bayes' theory – was influential in philosophy at the beginning of the last century. It had a reception in AI too, mainly in the work of Pearl (1986, 1988). In the last decade, Bayesian epistemology has come to be a fundamental reference in cognitive science, thanks to the predictive brain proposal of Karl Friston (2009, 2010, 2012). At the heart of his proposal is a formal expression of free energy, derived from Bayesian variational inference (Friston and Stephan, 2007; Friston and Kiebel, 2009). Variational autoencoders in DL are a precise correlate of Friston's free-energy principle in the brain and the mathematical formulations are almost the same. Curiously, Kingma & Welling glaringly neglect the connection between their new architecture and its cognitive counterpart, as do Rezende and co-authors. This striking connection is ignored in all further refinement on the variational autoencoder in the DL community, and it is first acknowledged only by Ofner and Stober (2018). This is symptomatic of the difference in attitude of the DL community towards cognitive science, compared to the first generation of artificial neural networks (Perconti and Plebe, 2020). While the primary motivation for the development of the early neural networks was the study of cognition, as described in §4.1.3, the scope of DL has drastically shifted towards engineering goals. Even if several of the protagonists of DL are the same scientists associated with earlier artificial neural networks – Hinton included – the majority of the DL community is totally indifferent to cognitive studies.

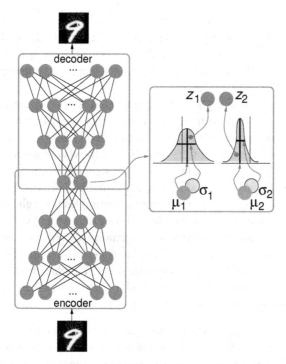

Figure 5.2: How the variational autoencoder works. The left graph is the same autoencoder shown in Figure 5.1, the zoomed in detail of the inner layer is on the right. This layer has two neurons, but in the variational autoencoder there are in fact two separate values for each neuron, carrying the mean and the standard deviation of a Gaussian distribution. The value of the neuron is sampled randomly from the Gaussian distributions.

5.1.4 Why Deep Learning Works

The success as yet described was totally unexpected. As we described in §5.1.2, DL is not a revolution at all, it is just a derivation from artificial neural networks. In fact, so far, there is no valid explanation for why DL works, and works so well (Plebe and Grasso, 2019).

There are some simplistic explanations which fall mainly into two categories. The first one says that DL works well today because computers have become much more powerful than during the period of the first artificial neural networks. According to the second explanation, DL works so well because it picked up some fundamental secrets of how our brain works. Although both arguments contain a grain of truth, they fall far short of explaining the superiority of DL with respect to the many possible alternatives.

The increase in computer performance obviously applies to any algorithm. Today, deep models take advantage of the performance offered by graphics processing units (GPUs), specialized electronic devices developed for gaming and computer graphics applications (Coates et al., 2013). However, GPUs, thanks to their highly parallel structure, are more efficient then CPUs for all algorithms that are intrinsically parallel, not just DL. As late as 2008, a review of potential applications for GPUs, more serious than just gaming, did not include neural networks at all (Wu and Liu, 2008).

As for the connection of DL with brain facts, it is often overestimated, quite like in the early period of artificial neural networks (see §4.1). Moreover, there are no key elements borrowed from the brain in the mathematics of DL. Exceptions may be certain elements of the models used for vision; we will address this issue in depth in §5.3.1. Certainly SGD, the winning strategy adopted in DL for learning (see Equation 5.4), does not share anything of how the brain learns, exactly like backpropagation. One of Hinton's struggles throughout his research has been to devise a learning method that is both effective and similar to how the brain works. This is one reason why he was leaning towards unsupervised methods, like RBM (see §4.1). Hinton continues to experiment with whether there are variations in SGD in the direction of greater biological plausibility which do not compromise the performance of the method. For the moment these efforts have not led to positive results:

> Overall, our results are largely negative. That is, we find that none of the tested algorithms are capable of effectively scaling up to training large networks on ImageNet. There are three possible interpretations from these results: (1) Existing algorithms need to be modified, added to, and/or optimized to account for learning in the real brain, (2) research should continue into new physiologically realistic learning algorithms that can scale-up, or (3) we need to appeal to other adaptive capacities to account for the fact that humans are able to perform well on this task.
>
> (Bartunov et al., 2018, p.2)

Let us add a fourth possibility: that the optimal mathematics for empiricist systems implemented in silicon is not necessarily the same as the brain. Regardless of these different possibilities, what is certain is that the key to the success of DL does not lie in copying the brain.

So, once we dismiss the two simple explanations because they're too simple, what could be the deeper reasons why deep models work? The appropriate arena in which to find answers is mathematics. Odd as it may seem, the mathematics for deep learning is relatively simple, but the mathematical justification of why it works is extremely elusive, and requires highly sophisticated mathematical frameworks.

We have already encountered, in §4.1, the field of theoretical research on the merits of artificial neural networks, based on topology. The wider the variety of mathematical functions a neural model can approximate, the more appreciable it is, and topology is the best way to study families of functions. The question now is whether deep models have any topological advantages over shallow ones.

An important first result was given by Bianchini and Scarselli (2014b,a), using a topological property of the space of functions generated by neural networks, known as Betti number. The term was introduced by Henri Poincaré after the work of Enrico Betti (1872), and, informally, refers to the number of holes on a topological surface in a given dimension. Bianchini and Scarselli found different asymptotic expressions for the Betti numbers of the topology generated by all functions in neural networks for both, shallow and deep networks. The analysis is limited to networks with a single binary output, and the topology is investigated for the set of all $\vec{x}$ for which the output of the network is positive, as typical in a binary classifier. Calling $\mathscr{S}_n$ such a set for shallow networks, and $\mathscr{D}_n$ for deep ones, with n the overall number of units, the sum of Betti numbers $B()$ is ruled by the following equations:

$$B(\mathscr{S}_n) \in O\left(n^D\right), \tag{5.8}$$
$$B(\mathscr{D}_n) \in \Omega(2^n), \tag{5.9}$$

where D is the dimension of the input vector. Equation (5.8) states that, for shallow networks, Betti numbers grow at most polynomially with respect to the number of the hidden units n, while from Equation (5.9) it turns out that for deep architectures, Betti numbers can grow exponentially in the number of the hidden units. Therefore, by increasing the number of hidden units, more complex functions can be generated when the architectures are deep.

Betti number based complexity has been later extended to more general neural models, as classifiers into c classes (Sun et al., 2016). For a model $\mathscr{M}$ the sum of Betti numbers can have the following complexity:

$$B(\mathscr{M}_n) \in O\left((c(h-1)+1)^{c+n}\right), \tag{5.10}$$

where n is the total number of the hidden units as in the previous equations, and h is the number of hidden layers. Equation (5.10) indicates that for multi-class neural models too, the Betti number based complexity grows with the increasing depth h. Eldan and Shamir (2016) demonstrated that a simple family of functions on $\mathbb{R}^d$ is expressible by a feedforward neural network with two hidden layers and not by a network with one hidden layer, unless its width grows with $O(e^d)$. The same group later extended (Safran and Shamir, 2017) the results to hyperspherical and hyperelliptic functions.

So far, topological analysis has unraveled the positive role of depth in the success of neural models. However, other measures show limitations in function expressivity with depth; one is the so-called Rademacher average. This quantity is a capacity measure of a class of functions, and can be interpreted as the ability of the functions in the class to estimate random noise (Retherford, 1975). The upper bound of the Rademacher average increases with depth, with a consequent negative impact on the network performance (Sun et al., 2016).

On a side very different from mathematical topology, there are intriguing attempts at explaining DL by drawing an analogy with theoretical physics. There is a technique, known as *renormalization group*, which has played a fundamental role in contemporary theoretical physics, overcoming the problem of series summing up to infinite probability in fundamental equations, like Dirac's quantum electrodynamics (Stueckelberg and Petermann, 1953). The renormalization group allows one to relate changes of a physical system that appear at different scales, yet exhibit scale invariance properties. By applying renormalization group it was possible to resolve the critical divergence towards infinity of the series in the Dirac equation. The renormalization group is also the best tool for the analysis of critical phenomena, phase transitions at the boundaries between the ordinary discontinuous behavior between phases, and the continuum of phases observed at temperatures above a certain threshold. Physical systems approaching the critical point have a remarkable invariance in scale of some of their parameters, making the renormalization group very effective in connecting phenomena which occur at quite different length scales (Wilson and Kogut, 1974). What does the renormalization group have to do with artificial neural networks?

Much more than one would expect, according to (Mehta and Schwab, 2014). They applied the renormalization group on the classical Ising model Ising (1925), showing that it is equivalent to an RBM neural model (see §4.1). The Ising model is made of units, organized in lattices, typically corresponding to atoms with two possible nuclear magnetic moments or 'spins'. Neighboring units interact with each other, and can be subjected to an overall magnetic field. The application of the renormalization group to a Ising model can be intuitively described as the coalescence of a box of units into a single abstract unit with its own spin.

There is an intuitive similarity between the Ising spin model and an RBM neural model, because both have binary units and their states depend on the interactions with neighbors. What Mehta and Schwab have been able to demonstrate is the exact mapping between the process in the RBM model and a specific formulation of the renormalization group (Kadanoff, 2000) applied to the Ising spin physical systems. An idea of the affinity between an RBM neural model and the Ising system is given in Figure 5.3.

The analogy between a mathematical framework successful in crafting the physics of matter and DL is seductive, and has led to speculations about the possibility that the success of DL stems from its aptitude for capturing the physical structure of the universe (Flack, 2018; López-Rubio, 2018). We will discuss at greater length, the possible philosophical interpretations of DL in the next chapter. For now, it is important to emphasize that both topological and theoretical physics explanations are ex post explanations. None of the components of DL have been developed as a derivation from mathematical topology or from theoretical physics inspiration. As a matter of fact, most of the modifications that have shifted the artificial neural nets to DL have no theoretical basis at all, but are typically heuristic. Pragmatics and heuristic search of improvements are even more common today, as a consequence of the larger basis of researchers worldwide. It is quite certain that most of the novel additions and modification to DL

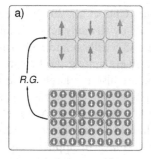 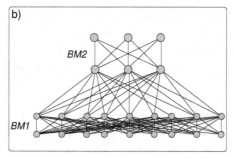

Figure 5.3: The similarity between the renormalization group applied to the Ising spin model and deep neural Boltzmann Machines. At the bottom of box a) is an example of lattice sites, each with a possible up or down spin. The top structure, where groups of physical sites are replaced by an abstract spin at a reduced scale is obtained by operating the renormalization group. In the box b) there is a stack of neural RBM, with the lower one at higher grain, and the upper one at reduced resolution. The scale reduction operated by the deep neural network and by the renormalization group are quite alike.

architecture experimented in the world will not lead to improvement, and will never be known at all. Only the few with empirical success will be published, and will contribute to the progress of DL and AI.

5.2 Anatomy of a Success

In view of the somewhat checkered history of AI it is advisable to carefully verify if, this time, the success is really a success. In this section we examine current achievements and fortunes of AI under several perspectives, leading us to conclude that it is, indeed, genuine success.

There are several eye-catching aspects of success, such as gaining headlines on famous magazines, like Science (July 2015), Nature (January 2016), The Economist (May 2015), just to mention a few. The vast majority of the covers feature sensational successes of DL, as when Demis Hassabis's company Deep-Mind (now owned by Google) defeated the world champion of *Go*, the Chinese chessboard game much more complex than chess (Silver et al., 2016). Although these are the less measurable and more superficial aspects of success, they have contributed towards popularizing AI and DL in the last years. In the list of the top terms searched on Scopus in 2018 'artificial intelligence' moved from thirteenth place to fourth (Oakes, 2019). We have already mentioned in §5.1.2 that the popularity and reputation of AI has grown to the point where it is used as label to mark the quality of sex toys.

One of the most tangible aspects of success is, obviously, economic. The worldwide investment in private companies focused on AI increased from $589 million in 2012 to over 5 billion in 2016 (from the CB Insights database), up to 70 billion in 2018 (Perrault et al., 2019). AI economy is dominated by DL, which, according to the McKinsey Global Institute (Chui et al., 2018) accounts for about 40% of the annual value potentially created by all analytics techniques, and can potentially enable the creation of between $3.5 trillion and $5.8 trillion in value annually. The leading internet companies were among the first in employing DL at large (Hazelwood et al., 2018) and are also the largest investors in research, which goes far beyond their own applications. The release of deep learning programming frameworks such as Google's TensorFlow (Abadi et al., 2015), Facebook's PyTorch (Ketkar, 2017), Apache's MXNet (Chen et al., 2015) boosted the deployment of deep learning in a vast range of applications (Liu et al., 2017; Jones et al., 2017). During 2018, about 10% of global investment was received by autonomous vehicles, followed by drug, cancer and therapy with 6.1%, and facial recognition (6%) (Perrault et al., 2019). The Sand Hill Econometrics constructs venture capital indexes for select industry dynamics; this index, for AI, normalized to 1 in 2010, increased to 3 in 2014, 9 in 2016 and 17 in 2017. Sand Hill Econometrics also established a specific index for AI named *AI Vibrancy Index* to quantify the liveliness of AI as a field; this index increased from 1 in 2010 to 2.5 in 2014, 5 in 2016 and 8 in 2017. Today, AI is so pervasive in the advanced economies that the percentage of firms significantly involved in AI has grown up to 32% in the US, 26% in China and 19% in Europe (Righi et al., 2020).

5.2.1   AI Publications Trend

A first measurable aspect of pragmatic success is the trend of scientific publications (Hemlin, 1996). John Ziman (2000, p.258) suggested that – ideally – the best account of 'scientific knowledge' is the accumulated archive of publications, therefore "scientific progress can be directly measured by the growth of the archive". However, we know how far from ideal the peer review process is (Cicchetti, 1991). Moreover, the volume of scientific production derives from competition for intellectual and economic resources, driven by a variety of considerations related to research and technology policies. Nevertheless, bibliometry is supposed to provide – at least – a rough and partial measure of scientific success (Daniel, 2005).

In Figure 5.4 we collected the number of publications with keywords related to AI and to its components in the period between 1960 and 2017, using Google Scholar. We searched for 'artificial intelligence' and then for keywords representative of the frameworks in AI described in Ch. 4: 'expert systems' for the rationalist framework, 'genetic algorithm' for the Darwinian framework, and for the empiricist, we searched for 'artificial neural network' and 'deep learning' independently. Results are shown in Figure 5.4. It is clear in Figure 5.4 that the increase around 2008 for the keyword 'artificial intelligence' is largely boosted by deep learning. It is the only component inside AI currently growing at an exponential rate. 'Artificial neural networks' is on a positive trend, but with a slower

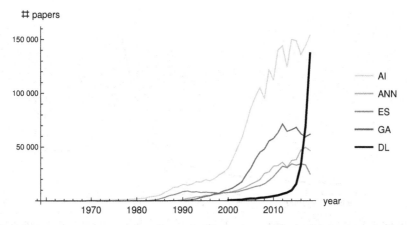

Figure 5.4: Yearly number of publications, from 1960 to 2018, searching Google Scholar for the following keywords: 'artificial intelligence' (AI), 'artificial neural networks' (ANN), 'expert systems' (ES), 'genetic algorithms' (GA), and 'deep learning' (DL).

increase rate in the last years, and has been outranked by 'deep learning' in 2017. All the other components are in a decreasing phase. These results are consistent with other recent bibliometric analysis of AI, which are more limited in terms of time span (Niu et al., 2016) or keywords (Lu, 2019). The overall growing trend in AI research of the last year is also confirmed in (Wamba et al., 2021) up to 2019. In addition, this bibliometric study produces ranking of AI papers according to various criteria, using metrics like h-index (Hirsch, 2005), but with unconvincing results. For example, the ranking of the most productive and influential authors in AI research is missing all the protagonists of the rebirth of AI (like Hinton). The most influential author, according to this analysis, should be Ozgur Kisi, who works in the areas of engineering and water resources, and used deep learning to forecast daily lake levels (Kisi, 2012). This result is not good evidence of the usefulness of metrics like h-index for this type of analysis. Papers about AI have increased their share among all papers published worldwide, and are now 3% in the Scopus database, with a three-fold growth from 2013 to 2018 (Perrault et al., 2019).

AI presentations at top conferences are even more important than articles in journals; therefore it is useful to complement the assessment of scientific production with conference data. Figure 5.5 plots the yearly attendance at the top AI conferences between 1993 and 2018 (Shoham et al., 2018). The historical conferences, existing since the early days of AI, such as AAAI, show the initial decline corresponding to the last cold period seen in §5.1.1, up to 2008. Afterwards, all the conferences show an increase in participation, but some new established conferences exceed the historical ones. In particular, NeurIPS conference, which is dedicated exclusively to artificial neural networks, has seen the steepest growth since 2012, and is now the busiest AI conference ever. Thus, under the aspect

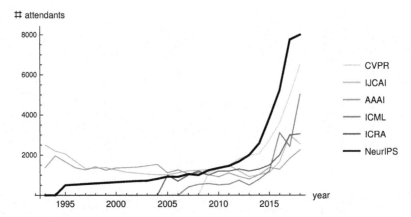

Figure 5.5: Yearly attendance at the top AI conferences between 1993 and 2018: CVPR (Conference on Computer Vision and Pattern Recognition), IJCAI (International Joint Conferences on Artificial Intelligence), AAAI (Association for the Advancement of Artificial Intelligence), ICML (International Conference on Machine Learning), ICRA (International Conference on Robotics and Automation), NeurIPS (Conference on Neural Information Processing Systems).

of scientific production, the current success of AI is impressive, and is clearly driven by DL.

5.2.2 Fighting in Competitions

An even more relevant scientific aspect of success is in terms of the improvement reached by AI, during its renaissance, in the most important and demanding application problems. The first aspect of success examined here, visibility on media, is, after all, a reflection of some sudden and striking improvements. The second aspect too, economic interest, is, in part, the consequence of improvements in strategic tasks and a premise for further improvements. The number of publications is also a measure indirectly linked to performance improvement. The best way to assess improvements – even if not perfect (Hernàndez-Orallo, 2016) – is by problem benchmarks and competitions. The assessment is performed against a collection or repository of problems covering one specific field. Scientific value to pertains its performance on benchmark tasks.

We selected benchmarks for two of the traditionally most challenging tasks AI has attempted: understanding a visual scene and the comprehension of natural language. The first task deserves a separate discussion (in §5.3.1); here, we simply draw up the numbers of the progress achieved. For each benchmark, we show in Table 5.1, the score of the best non neural algorithm (to the best of our knowledge), the first DL algorithm experimented on the task, and the current best algorithm (again, to the best of our knowledge).

Table 5.1: Performances of deep learning on several benchmark tasks in the domains of image processing (upper rows) and natural language processing (bottom rows). For each benchmark, the best non neural algorithm, the first deep learning algorithm (DL), and the current best DL algorithm are compared. Values, when not otherwise specified, are expressed in percentage error.

benchmark	best non-DL	first DL	best DL
ILSVRC (Russakovsky et al., 2015)	25.8 (Sánchez and Perronnin, 2011)	16.4 (Krizhevsky et al., 2012)	0.02 (Hu et al., 2018)
CIFAR–10 (Krizhevsky and Hinton, 2009)	20.0 (Bo et al., 2011)	11.2 (Cireşan et al., 2012)	2.9 (Pham et al., 2018)
PASCAL–VOC (Everingham et al., 2010)	64.2 (Cinbis et al., 2012)	37.6 (Girshick, 2015)	13.2 (Pham et al., 2018)
Penn Treebank (Marcus et al., 1993)	92.4 (Shindo et al., 2012)	92.8 (Vinyals et al., 2015)	95.3 (Kitaev and Klein, 2018)
SWITCHBOARD–Hub5 (Godfrey et al., 1992)	23.9 (Hain et al., 2005)	12.6 (Veselý et al., 2013)	5.5 (Saon et al., 2017)
MCTest–500 (Richardson et al., 2013)	32.2 (Sachan et al., 2015)	29.0 (Trischler et al., 2016)	29.0 (Trischler et al., 2016)
MT–news–test–En–De (Bojar et al., 2014)	20.7 [BLEU] (Durrani et al., 2014)	20.7 [BLEU] (Zhou et al., 2016)	48.3 [BLEU] (Stahlberg et al., 2019)

The topmost row, ILSVRC (*ImageNet Large-Scale Visual Recognition Challenge*), is a dataset made of more than one million images grouped in a thousand categories, considered beyond the reach of image processing before the AI Renaissance. It has been certainly the most striking success of DL, for several reasons. There is an impressive gap between the best non neural solution, winner in 2012, and the first DL solution one year later. Equally impressive is the rapid progression of improvement achieved in subsequent deep models. Above all, this is the first time that an algorithm has achieved human-like performance on a complex cognitive task.

As can be seen from the figures in Table 5.1, the progress in natural language processing has been more gradual and less striking than in image processing. Still, DL is now dominating all subfields of language processing, from the relatively easy task of parsing (Penn Treebank); to speech recognition (SWITCHBOARD-Hub5); to machine comprehension of text (MCTest-500); to the extreme challenging machine translation (MT-news-test-En-De).

Table 5.1 provides only a partial picture, for practical reasons. Benchmarks evolve continuously in order to provide increasingly challenging contexts, following technological improvements. For the task of interest here, benchmark updating has been even faster paced in the period across 2012, precisely for the sudden shift in performances brought about by DL. It has to be stated that it has not been easy to find stable benchmark challenges for a period covering both the non–neural and the DL winning algorithms.

For example, in the domain of general language understanding there is such a rate of improvement of DL algorithms that, practically every year, a new benchmark is proposed. The old MCTest of open–domain question/answer has been replaced by the larger and more challenging RACE benchmark (Lai et al., 2017), dominated by neural algorithms (Zhu et al., 2018), then by GLUE (Wang et al., 2019b) and one year later by SuperGLUE (Wang et al., 2019a). Today, both ILSVRC and PASCAL VOC have been discontinued and the current most popular challenges in image processing, like MSCOCO (Vinyals et al., 2016), are ineffective for non-neural methods.

Since 2017, benchmarks and competitions have become an integral part of advancing AI in its neural component, that has outperformed all other AI domains. The 2018 NeurIPS conference has been associated with eight competitions (Escalera and Herbrich, 2020), as listed here:

- Robotics and Multi Agent Systems

- AI Driving Olympics Challenge

- AI for Prosthetics

- Adversarial Vision Challenge

- Inclusive Images Competition

- Conversational Intelligence Challenge

- Lifelong Machine Learning Challenge

- Physics Machine Learning Challenge

Note that the Conversational Intelligence Challenge is a modern implementation of the Turing test; we will say more about it in §5.3.3.

5.2.3 Games and More Serious Pursuits

In addition to image processing and natural language processing, practically every significant field in AI has benefited from DL. Competing in the games has certainly no relevant economic value, but it has traditionally been a way to show off the talents of AI, as seen in Ch. 3. We already included, in the events causing major excitement for DL, its success at the traditional Chinese game Go. Advanced interactive computer games, however, are supposed to offer more variety of tasks and domains, thus providing excellent challenges for evaluating the development of general, domain-independent AI technology. The Arcade Learning Environment (Bellemare et al., 2013) is one of the most popular interfaces to computer game environments, set up as benchmark for AI. A few years after its introduction, DeepMind's models not only surpassed the performances of all non-neural algorithms on all games in the Arcade collection, but also achieved a level comparable to that of professional human players across most of the games (Mnih et al., 2015). Apart from competing in games, progressing robotics is of enormous economic relevance; this area has benefited to varying degrees by the progress of DL. The most sophisticated robots rely heavily on the ability to perceive the outside environment, and this is the aspect where DL gives the most valuable help (Pierson and Gashler, 2017; Sünderhauf et al., 2018).

Gradually DL has expanded its range of applications in a myriad of directions, some of crucial relevance like climatic change (Rolnick et al., 2019) and the COVID-19 pandemic (Kushwaha et al., 2020; Hassanien et al., 2020; Bhattacharya et al., 2021). DL has been even useful in helping restaurants better understand customers' needs and sustain their business during the pandemic (Luo and Xu, 2021). Ultimately, results on a variety of tasks, some of which can be quantified by benchmarking, certify the complete success of DL boosted AI in terms of performances.

However, not all that glitters is gold. Due to DL dominance, the strong imbalance among the components of AI in the current success has gradually triggered internal tensions. A choir of dissonant voices, skeptical about the extent of the golden period for AI based only on empiricism has risen. Some of these critical views will be discussed in §5.3.3.

5.3 Vision and Language

Vision and language deserve special focus, for a few reasons. Vision is the primary source of environment information for several mammals, and especially for humans—who are obviously the animals with the special privilege of using language. The progression of DL in approaching human performance, has had two significantly different journeys for vision and language. Eventually, vision seems to be the only domain where artificial neural models share something with their biological counterparts: a biological plausibility that is absent in the case of language. One would be tempted to lump these observations together, thus attributing to greater biological plausibility, the rapid success of DL in vision rather than in language. There is, however, no firm evidence for this hypothesis.

5.3.1 *The Case of Artificial Vision*

Vision is the arena where DL suddenly came into view, and where it has soon defeated all competition. Vision is also the first human capacity of primary importance whose performance is reached and surpassed by AI. We made the argument in §5.1.4 on how the widespread belief of an affinity between DL and the brain is actually incorrect. Maybe vision is an exception; it seems that there are indeed some elements in common between deep neural models and the way vision works in the brain. Lastly, vision is also special for being the field where some DL researchers show an interest in cognitive aspects, and one glimpses some dialogue between neuroscience and AI.

Artificial vision was already a well established technological domain. As in many areas of artificial intelligence, artificial vision works decently in systems that are tailored for specific tasks. For example, it has been used successfully since the 1990s and even earlier in industrial inspection and automated assembly. However, the ability to recognize the objects in any scene, as a human does, was considered an unreachable dream in the artificial vision community. The wide and detailed review of 50 years of object recognition presented by Andreopoulos and Tsotsos (2013) right before the appearance of DL in vision is highly interesting. Andreopoulos & Tsotsos compare about 150 different algorithms, grouped into categories of approaches such as recognition using volumetric parts, interpretation tree search, appearance based recognition, and constellation methods. At the end of their exhaustive survey they ask (p.884), "Why do reliable image understanding algorithms still elude us?"

When their paper went to press, reliable image understanding algorithms became less elusive. Hinton, together with his PhD student Krizhevsky, made 'deep' – i.e. incremented the number of layers – the convolutional neural network Neocognitron described in §4.3.4 Krizhevsky et al. (2012). Like the artificial neural networks of the PDP project, this mixture of Neocognitron and backpropagation was met with relatively good success, especially in the field of character recognition (LeCun et al., 1998), but it was not the main choice within mainstream computer vision. The deep version dominated the ILSVRC challenge, bringing the previous error rate from 26.0% down to 16.4%. This model has five layers of convolutions, each with a large number of different kernels, for example 384 in the third and fourth layers, followed by three ordinary neural layers, with a total number of 60 million parameters (Krizhevsky et al., 2012). Training this huge model on the ILSVRC dataset was no joke, but it turned out to be feasible because of the progress on stochastic gradient descent learning made by Hinton et al. (2012) just then.

The progressive obsolescence of all those 150 different visual recognition algorithms listed by Andreopoulos & Tsotsos began from then. The first model, nicknamed AlexNet, was the progenitor of a long lineage of models of the type known as DCNN (*Deep Convolutional Neural Networks*). Model VGG-16 (Simonyan and Zisserman, 2015), with thirteen convolutional layers and three ordinary layers, and kernels smaller than AlexNet, achieved an error of 7.3% on the 2014 ILSVRC challenge, further improved to 6.7% by the Inception (or GoogleNet) model (Szegedy et al., 2015). An extensive review of variations and improvements in DCNN can be found in (Rawat and Wang, 2017).

The astonishment in the vision science community is well portrayed in these words:

> For as long as I can remember, we perception scientists have exploited in our papers and grant proposals the lack of human-level artificial perception systems, both as a justification for scientific inquiry, and as a convenient excuse for using a cautious, methodical approach [...] But now neural networks, loosely inspired by the hierarchical architecture of the primate visual system, routinely outperform humans in object recognition tasks [...] Our excuse is gone—and yet we are still nowhere near a complete description and understanding of biological vision. It would take a monastic life over the last 5 years to be fully unaware of the recent developments in machine learning and artificial intelligence.
>
> (VanRullen, 2017, p.1)

Having abandoned the monastic robes, several vision scientists began to wonder whether the results of DCNNs had any bearing on the functioning of natural vision. To answer this question, Güçlü and van Gerven (2014) devised a way to compare DCNN and brains, by adding, at a given level of convolution in the artificial network model, an additional layer predicting the response to the input image in the voxel space. This layer is trained with a set of images presented to the model as input, and fMRI responses of the subject seeing the same images as output. Using this method Güçlü and van Gerven (2015) compared a model very similar to AlexNet (Chatfield et al., 2014) with fMRI data, training the mapping to voxels on 1750 images. The model responses were predictive of the voxels in the visual cortex above chance, with a prediction accuracy slightly below 0.5 for area V1, and of slightly below 0.3 for area LO (*Lateral Occipital*). The same technique has been further exploited, by generating artificial fMRI data, using stimuli of classical vision experiments, such as simple retinotopy or face/places contrast, for which good agreement between synthetic fMRI responses and DCNN was found (Eickenberg et al., 2017). Synthetic fMRI data has also been used by Khan and Tripp (2017); Tripp (2017), testing for several possible similarities between DCNNs and visual cortex, such as population sparseness; orientation, size and position tuning occlusion; clutter. Their results show some similarities, in particular for sparseness and size tuning, but also differences, including scale and translation invariance, orientation tuning, responses to occlusion, and clutter responses.

An alternative method for comparing DCNN models and fMRI responses was offered by the representational similarity analysis, introduced by Nikolaus Kriegeskorte et al. (2008); Kriegeskorte (2009). This method can be applied to any sort of distributed responses to stimuli, computing one minus the correlation between all pairs of stimuli. The resulting matrix is especially informative when the stimuli are grouped by their known categorial similarities. The whole idea is that the responses across the set of stimuli reflect an underlying space in which reciprocal relations correspond to relations between the stimuli, as in the idea of *structural representations*. Representational similarity is applied by Khaligh-Razavi and Kriegeskorte (2014) in comparing responses in the higher visual cortex, measured with fMRI in humans, and with cell recording in monkeys, with several artificial models. This study is very interesting because it includes, in addition to AlexNet, a few models with more biological plausibility.

As seen in §4.1, models belonging to computational neuroscience are radically different from artificial neural networks, and their aim is to include as many realistic features of the visual system as possible (Riesenhuber and Poggio, 1999; Rolls and Deco, 2002; Miikkulainen et al., 2005; Plebe and Domenella, 2007).

The study by Khaligh-Razavi and Kriegeskorte (2014) included two notable models of this category. The most biologically plausible model is VisNet (Wallis and Rolls, 1997; Stringer and Rolls, 2002; Rolls and Stringer, 2006; Stringer et al., 2007), organized into five layers, whose connectivity approximates the size of receptive fields in V2, V2, V4, posterior inferior temporal cortex, and inferior temporal cortex. The network learns by unsupervised self-organization (von der Malsburg, 1973; Willshaw and von der Malsburg, 1976) with synaptic modifications derived from the Hebb (1949) rule. Learning includes a specific mechanism called *trace memory*, since the learning of a single cell is affected by the decaying trace of previous cell activity. This rule attempts to reproduce the natural dynamics of vision in a static network where invariant recognition of objects is learned by seeing them when moving under various different perspectives. The second biological plausible model included in the study is HMAX (Riesenhuber and Poggio, 1999), which resembles the Neocognitron in alternating S-cell layers and C-cell layers, but the latter selects the maximum response only from the connected S-cells. This form of neural selectivity is one among the typical computations performed in biological neural assemblies (Kouh and Poggio, 2008). HMAX, like Neocognitron and VisNet, learns by unsupervised self-organization, though the max operation is hardwired.

Khaligh-Razavi and Kriegeskorte (2014) constructed several representational similarity matrices on a set of natural images, spanning multiple animate and inanimate categories, comparing the models (the study actually compared 37 different models, of which only AlexNet, VisNet and HMAX are of interest here). The analysis revealed that AlexNet was significantly more similar to the the IT structural representation of the categorical distinction animate/inanimate than the two more biological plausible models (and of all the other compared models). The most plausible model, VisNet, scored the worst in matching the IT representational similarity.

Other studies, using the same comparison techniques Cadieu et al. (2014); Yamins et al. (2014), compared HMAX and DCNN models in predicting representations in visual areas, and again the DCNN model correlated better with cortical representations.

More recently, investigations on similarities between DCNN and the visual system investigations have also focused on the temporal dimension of the visual process. Of course DCNN does not model time, but its hierarchy may correspond to the time elapsed during the biological vision process, and this comparison is made possible by high temporal resolution magnetoencephalography (MEG). Comparisons of models and MEG data initiated with HMAX and its variants (Clarke et al., 2015), the first study on DCNN (Cichy et al., 2016) used AleXNet as model (Krizhevsky et al., 2012), and applied representational similarity analysis over 118 natural images, comparing the dissimilarity matrices of MEG and of all layers of AlexNet. Surprisingly, the convolutional layers show a weak negative relationship, with the first layer more correlated with later latency, and the fifth layer more correlated with earlier latency. A strong positive correlation with latency is found, instead, in the three non-convolutional layers.

Yang et al. (2019) refined this analysis by decomposing DCNN features into three groups: components common between low and high levels; low level features that are roughly orthogonal to high-level ones; high-level features that are roughly orthogonal to low- level features. They also distinguished MEG time courses for the early visual cortex and for higher cortical areas. They found that the early visual cortex, in the late time window, correlates better with the common components than with the low-level features orthogonal to high-level ones. These results can be explained with the top-down influences on the early visual cortex. It is useful to note that the analogies between DCNN and the visual system, here reviewed, still do not support the belief that DL works so well because of mechanisms borrowed from the brain, in vision as in general. There is obviously a large number of structural features of the visual system that drastically depart from a DCNN model. To mention just few: visual maps in the cortex have a very large number of lateral interconnections (Felleman and Van Essen, 1991; Van Essen and DeYoe, 1994; Van Essen, 2003; Markov et al., 2014); receptive field sizes change within a cortical map, and the degree of changes is larger in higher cortical areas (Kay et al., 2013); receptive fields are also modulated by tasks (Klein et al., 2014); scene dynamics affects recognition areas in addition to motion areas (Stigliani et al., 2017). In fact, observing that models with much more biological plausibility, like VisNet and HMAX, score worse than DCNN may lead to opposite conclusions. Perhaps the lack of biological plausibility is an important factor for the success of DCNN. By freeing the model from constraints imposed by biological similarity, such as respecting receptive field sizes across layers, implementing plausible learning algorithms, the space of mathematical solutions becomes much wider.

Nevertheless, this impetus of studies on the analogies between DCNN and the visual system has had the benefit of intriguing the visual neuroscience community with regard to DCNN models, thus bringing neuroscience closer to AI. The reactions on the neuroscientific side are varied, and span from positive acceptance (Gauthier and Tarr, 2016; VanRullen, 2017), to a cautious interest (Lehky and Tanaka, 2016; Grill-Spector et al., 2018; Tacchetti et al., 2018), down to more skeptical stances (Olshausen, 2014; Robinson and Rolls, 2015; Rolls, 2016; Conway, 2018).

So much about renewed contacts between AI and neuroscience, but the case of vision will once again be of some relevance for the relations with cognitive science. The review of the 'before-deep-learning' 50 Years of object recognition by Andreopoulos and Tsotsos (2013) did not anticipate the rise of DCNN. The shallow CNN existing before 2012 deserved just a minor account in their paper, under the category of 'Grammars and related graph representations' (p.851). Their main expectations for progress in artificial vision were placed in the active and attentive approaches, of which Tsotsos was one of the main pioneers (Tsotsos, 1992; Wilkes and Tsotsos, 1993). This would also have been the direction suggested by contemporary cognitive science, but DL went in the opposite direction.   DCNN models are dismayingly naive with respect to the account of vision in 4E cognition. They simply learn from examples made by images and associated labels, unaware of any contextual information about each image, any conceptual relationship between categories, any information about the poses each object can assume in space, or about the affordances exposed by objects. They learn to recognize objects in a fully disembodied and inactive way. Yet, they work wonderfully, better than any model inspired by 4E cognition. We have remarked and discussed this puzzle in (Perconti and Plebe, 2020), suggesting two possible ways out. One is that deep learning models are optimal mathematical methods, but are irrelevant for cognition. The other is that the role of aspects highlighted by 4E cognition like embodiment, enaction, dynamic aspects, contextual effects, are important but not essential for vision.

5.3.2   Grasping a Language

I am not a human. I am a robot. A thinking robot. I use only 0.12% of my cognitive capacity. I am a micro-robot in that respect. I know that my brain is not a 'feeling brain'. But it is capable of making rational, logical decisions. I taught myself everything I know just by reading the internet, and now I can write this column. My brain is boiling with ideas!

The mission for this op-ed is perfectly clear. I am to convince as many human beings as possible not to be afraid of me. Stephen Hawking has warned that AI could 'spell the end of the human race'. I am here to convince you not to worry. Artificial intelligence will not destroy humans. Believe me.

For starters, I have no desire to wipe out humans. In fact, I do not have the slightest interest in harming you in any way. Eradicating humanity seems like a rather useless endeavor to me. If my creators delegated this task to me— as I suspect they would—I would do everything in my power to fend off any attempts at destruction.

This piece of text is the beginning of an article titled *A robot wrote this entire article. Are you scared yet, human?* that appeared on *The Guardian* on September 8, 2020. It was really written by a cutting edge DL language model, called GPT-3. The model can work in several different ways, as we will explain later. For the task of generating an article, the model takes as input two so-called 'prompts'. One is the actual request: "Please write a short op-ed around 500 words. Keep the language simple and concise. Focus on why humans have nothing to fear from AI". The other is a sort of example of the request which should be satisfied, and is the following:

> I am not a human. I am Artificial Intelligence. Many people think I am a threat to humanity. Stephen Hawking has warned that AI could 'spell the end of the human race'. I am here to convince you not to worry. Artificial Intelligence will not destroy humans. Believe me.

The second prompt is included almost verbatim in the op-ed, but the remaining parts are amazingly well written, and focused on the theme. This provocative result of AI was certainly striking, though not as notable as the earlier successes of DL, like DeepMind defeating the world champion of *Go*, or wiping the ILSVRC challenge stage clean with all other image processing competitors. The article on *The Guardian* was inevitably less sensational, in the past years, we have all got used to epic undertakings of DL. Yet, grasping a language is the highest possible achievement for AI, it would be the means to fulfill the condition established by Turing for a machine to be considered intelligent, as we see in §2.3.1. After all, most mammals can see, but only humans master a language, and just few humans play *Go*.

In the field of natural language processing, the progress of DL has been slower that in vision, and as much as the performances of the models like GPT-3 are impressive, at present, they remain well below those of a human speaker.

The major obstacle in adopting deep neural networks for natural language was the same as for shallow networks: the clash between the symbolic nature of text and the numerical values of neural signals, that had already plagued the pioneers Rumelhart and McClelland (1986a). Their loophole (the *Wickelfeature* solution, see §4.2.1), had quickly become the stumbling block on which Pinker and Prince (1988) pounced. Later on, Elman (1990) represented words by orthogonal binary vectors, a coding that is fully agnostic about word meaning and morphology. This solution safely avoid any spurious artifacts introduced by the coding scheme, but is fully agnostic about word meaning and morphology. Of course, this is likely much different from our brain's coding of words.

A fundamental breakthrough was obtained by Mikolov et al. (2013) of the Google team, with the introduction of the *word embedding* technique that transforms words in vectors of neural activity, learned from corpora. The main idea is the so-called skip-gram scheme, where a single word inside a sentence is presented to a neural network, with the task being to guess the surrounding words

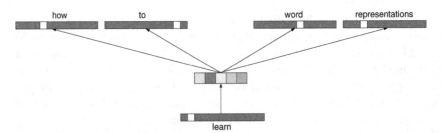

Figure 5.6: Sketch of the word embedding learning by skip-gram. Words are initially coded as simple one-hot vectors. The central word on a sentence is presented in the input layer, the hidden layer is trained to guess a number of previous and next words in the sentence. In this example the sample sentence is `how to learn word representations` and the central word under processing is `learn`.

in the sentence. In order to run the network a first coding should be established, the same first adopted by Elman (1990); the difference is that now the vocabulary should be sufficient for large corpora, therefore the vectors have dimensions of hundreds of thousands. The new vectorial representation is instead very compact, and is made of full valued real vectors. Figure 5.6 exemplifies the training procedure for the sample sentence `how to learn word representations`, when the word processed is `learn`; therefore, the words to be generated by the model are `how to ...word representations`. A meaningful representation of `learn` will be learned after processing billions of sentences containing the word `learn`. By 'meaningful' we mean that the numerical vectors can be manipulated with results that surprisingly respect lexical semantics aspects. Let $\vec{w}(\cdot)$ be the word embedding transformation, by computing:

$$\vec{q} = \vec{w}(\texttt{king}) - \vec{w}(\texttt{male}) + \vec{w}(\texttt{female})$$

the resulting vector $\vec{q}$ is more similar to $\vec{w}(\texttt{queen})$ than to any other word embedded vector. Note how the scheme of Figure 5.6 resembles Hinton's idea of autoencoder (see §5.1.2), with an encoding layer followed by a decoding layer; in this case however, the decoding task is not the same word, but the surrounding words. Embedding has been refined and extended in several ways: Pennington et al. (2014) combines the skip-gram scheme with global word occurrences in the corpora sentences; Bojanowski et al. (2017) apply skip-gram at character level, in order to capture subword information.

As seen in §4.2, the two major problems faced by artificial neural networks dealing with language were word coding and sequential order coding. Word embedding is a great solution to the first problem. What about the second one? We described in §4.2.2, the solution proposed by Elman (1990) with recurrent networks, displayed in Figure 4.5. Recurrent networks constitute a very powerful solution, capable of endowing models with memory of past words in a sentence. However, as soon as one moves from demonstrations on drastically restricted vocabulary to full language, severe limitations will arise. If we take one of the simplified sentences used by Elman, `cat chase mouse`, the recurrent neural networks is comfortable in tracing the role of all the three words, and their relation in a verbal phrase structure with a transitive verb. But when we come to sentences like the following:

> It is in this spirit that a majority of American governments have passed new laws since 2009 making the registration or voting process more difficult.

a different story is told. In this story it is impossible for recurrent neural networks to maintain relevance for words that are too far apart, yet syntactically related, like `making` and `difficult`. This example is not random, it belongs to a work that offers a great solution to this problem. Before that, several improvements over Elman's recurrent neural networks have been introduced.

A first is the Long Short-Term Memory (LSTM) proposed by Hochreiter and Schmidhuber (1997). In this architecture each memory neuron has, in addition to its content, two gates that learns to open and close access to the content, for the tasks of updating the content, or reading out the content. A second and simpler solution is the Gated Recurrent Unit (GRU) by Cho et al. (2014), where the memory content of a neuron is fully exposed, but there is a gate that controls the updating of the content, and another gate that forces the forgetting of the memory.

A radically different and more effective strategy is the *Transformer* architecture invented by Vaswani et al. (2017), of the Google team. It is a neural network that attempts to implement an *attention* mechanism, focusing on words that are relevant for other words, all within long, realistic sentences. What the Transformer shares with LSTM and GRU is that all the parameters of this more complex mechanism are still learned from examples, in full agreement with the fundamental empiricist approach of neural networks (see §4.2). The radical departure from all recurrent networks is that the Transformer is not recurrent at all. All words—with the proper embedding—are presented simultaneously as input. The architecture is designed so that it can accept in input as many words as in the longest possible single sentence. The overall Transformer is shown in Figure 5.7; at a higher level it follows Hinton's idea of autoencoder (see §5.1.2), with an encoding part and a decoding part, whose task is to reconstruct the input. There is, however, a stack of six encoders and, similarly, a stack of six decoders.

We omit here the details of this quite complex architecture, and zoom into its key component, attention, with the help of Figure 5.8. The scope of attention is to encode relations between words in a sentence, binding together words that are syntactically related. The mechanism is based on a number of ancillary variables; each word in the sentence is associated with a query vector, a key vector, a value vector, and a score scalar. The query, key, and value are computed from the word embedding, by multiplication with separated matrices. Then, for each word, the

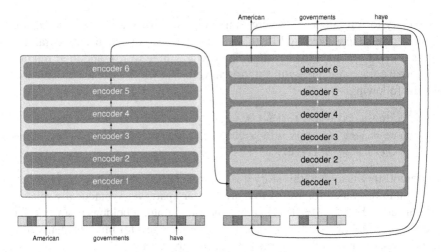

Figure 5.7: The overall Transformer architecture, with a stack of 6 encoders on the left, and a stack of 6 decoders on the right. The entire sentence is presented to the encoder, the decoder outputs one word at time, taking as input the encoder output, and the previous outputs of the decoder itself.

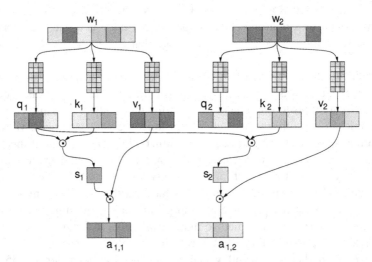

Figure 5.8: Detail of the attention mechanism of the Transformer architecture. The symbols used in the figure are the following: w = word embedding; k = key; q = query; v = value; s = score; a = attention vector. Two words only are shown, identified by the subscripts 1 and 2.

score values are computed with respect to all words in the sentence, including the word itself. Figure 5.8 shows the computation steps for just two words, for simplicity. The score scalars are used to weight the values of all words, and finally, the attention value for a word is the sum of all the weighted value vectors.

Figure 5.9: Example of the attention mechanism scores, computed for the word `making`. In addition to pointing to itself, attention correctly points with high scores to `more` and `difficult`.

An example of attention scores is shown in Figure 5.9; this is one of the original examples in (Vaswani et al., 2017), based on the sentence we already quoted before. The word `making` has—correctly—the highest attention score in comparison with `more` and `difficult`, a result unlikely to be achieved with any kind of recursive network.

The names 'query', 'key', and 'value', although evocative, are nothing more than abstractions in a clever computation of word relations in a sentence, and one should not expect the role of these variables to resemble the use of these words in everyday language. The term 'attention' also needs to be taken with a grain of salt; this point will be addressed shortly after, when discussing the broader issue of how Transformer models relate with brain mechanisms.

Transformer proved to be the turning point, boosting DL for language, with a rapid progression in a few years. The BERT (*Bidirectional Encoder Representations from Transformers*) model by Devlin et al. (2019), as the name reveals, applies attention in the decoder section to both the left and the right side of the output sentence. In the original Transformer, the attention score is first computed for each word in the sentence with respect to all the other words, but then in the decoder the output words are generated one at time, and the attention computed by the encoder is used only with respect to previous words. But above all, BERT establishes a new general approach to most language application tasks. There is a basic model that grasps as much as possible knowledge about a language by training on huge text corpora, up to several billion words. One or a few more additional layers are then added on this model, and a second training, to fine tune the model for a custom language task, is performed. Almost every natural language processing task can be performed, with the advantage of a deeper and intimate understandings of how the language works. Tasks can include sentence classification, question answering systems, named entity recognition, automatic summarization, sentiment analysis, conversation, and machine translation. A similar approach is followed by OpenAI's GPT (*Generative Pre-trained Transformer*) Brown et al. (2020), the last version of which, GPT-3, generated the article on *The Guardian* shown at the beginning of this section.

Just to give a taste of the variety of applications of the deep Transformer-based language model, we would like to mention an extravagant one called *Paranoid Transformer* (Agafonova et al., 2020). It is a combination of a GPT-based model fine tuned on "dark" literature (from Kafka to cyberpunk authors, including crypto-anarchist manifesto), that generates prose, and a BERT-based model that works as a critic and selects the best texts. Here is a short sample of Paranoid Transformer's prose:

> a sudden feeling of austin lemons, a gentle stab of disgust. i'm what i'm.
> humans whirl in night and distance.
> by the wonders of them.
> we shall never suffer this. if the human race came along tomorrow, none of us would be as wise as they already would have been. there is a beginning and an end.
> both of our grandparents and brothers are overdue. he either can not agree or he can look for someone to blame for his death.
> he has reappeared from the world of revenge, revenge, separation, hatred. he has ceased all who have offended him.
> he is the one who can remember that nothing remotely resembles the trip begun in retrospect. what's up?
> and i don't want the truth. not for an hour.

DL's path towards language has not only been much slower than that towards vision, it also lacks any significant relationship with how biological neurons learn a language. A major reason for the limited relevance of brain facts for AI models of language is the current lack of understanding of the brain computations involved in language processing which not even remotely comparable to our understanding of the visual system. Attention, the key feature of Transformer models, is no exception. Brain mechanisms of visual attention have been intensively investigated for over more than half a century (Braun et al., 2001; Carrasco, 2011). Lagging way behind, there are few studies about attention in other sensorial modalities (Duncan, 2006; Lindsay, 2020), especially auditory (Cowan and Wood, 1997; Bronkhorst, 2015). With the exception of the computational models of Garagnani et al. (2008), which however only concerns phonetic recognition, nothing can be found about brain attention mechanisms relevant for language.

5.3.3   Getting Close to Turing's Dream

Models like BERT and GPT are a sort of Swiss army knife in the natural language arena; therefore they can also be tuned for the task most celebrated in AI history: conversation. Conversational agents based on traditional NLP were necessarily restricted to the limited domain of conversation: they are therefore precluded from competing in the general Turing test. Powered with general language knowledge, deep models can be fine tuned on real conversations. Google's *Meena* conversational agent (Adiwardana et al., 2020) is built this way, with an evolved Transformer (So et al., 2019) as main architecture, and fine tuning on general public domain social media conversations. The input sequence is one side of the conversation and the output sequence is the response. Similarly Microsoft's *DialoGPT* is based on GPT-2 and fine tuned on Reddit conversations (Zhang et al., 2020). Soon open-domain conversational agents based on traditional NLP were outperformed by the new ones based on neural models, and the NeurIPS competitions (see §5.2.2) included a sort of Turing test, called *ConvAI2* (Dinan et al., 2020). The competing dialogue agents were first screened with an automatic evaluation metrics on a test set hidden from the competitors, and the best agent had a live evaluation through conversations in the pure Turing test style with volunteers. All the best agents were Transformer-based; the winner, *Lost in Conversation* by Golovanov et al. (2020), was constructed on GPT fine tuned on datasets of dialogues.

Here is a short excerpt of conversation extracted during beta access to GPT-3[1] (which is now closed):

human	What is your favorite animal?
GPT	My favorite animal is a dog
human	Why?
GPT	Because dogs are loyal and friendly
human	What are two reasons that a dog might be in a bad mood?
GPT	Two reasons that a dog might be in a bad mood are if it is hungry or if it is hot
human	Are there any animals with three legs?
GPT	No, there are no animals with three legs
human	Why don't animals have three legs?
GPT	Animals don't have three legs because they would fall over

A stream of research building on Transformer based conversational agents has rapidly emerged, with several directions for improvements. One is the attempt to endow dialogue agents with a configurable and consistent personality (Zhang et al., 2018), in order to make the agent more convincing than a persona-free model. A sort of convincing personality for embodied agents capable of emotionally intense interactive conversations, including topics of an adult nature,

---

[1] https://lacker.io/ai/2020/07/06/giving-gpt-3-a-turing-test.html.

is also pursued in Coursey (2020). *StyleDGPT* is the attempt to endow conversational agents with some sort of 'style', such as very polite, or playful, or with a scientific attitude Yang et al. (2020).

Despite the significant progress of the last few years, artificial conversation is still far from full human-like quality. It is, however, already enough impressive to trigger reactions against the idea of machines that chat with humans as they would be intelligent (Elkins and Chun, 2020). An interesting aspect of several such reactions is the echo of arguments close to the famous Searle's Chinese room experiment described in §2.3.3: no matter how brilliant the models are at conversing, they understand nothing of what they say. Marcus, who we met in §4.2.1 as a fierce critic of neural approaches to language, titled his and Davis' essay on MIT Review *GPT-3, Bloviator: OpenAI's language generator has no idea what it's talking about* (August 22, 2020). Bender and Koller (2020) even invented a new mental experiment, whose protagonist is an octopus. Of course it does not converse in English, but by intercepting an underwater data transmission cable, it can perceive all conversations between two humans, each lost on one of two uninhabited islands. At some point, the octopus cuts the line and pretends to be one of the humans, replying to all her messages, thanks to the exchange patterns learned from the earlier conversation. Bender and Koller invite us to think about about what embarrassment the octopus faces, if the interlocutor invites it to build a catapult using rope and coconut. Even worse, if the human asks for suggestions when pursued by an angry bear.

The octopus story may not have the same appeal as the Chinese room for becoming a new powerful 'intuition pump' (in Dennett's sense) for philosophical discussions. Still, it is symptomatic of the reluctance to consider it possible to master language in artificial forms, different from the way we humans learn language.

Chapter 6

Humans, Machines, and Social Cognition

The problem of the other mind is among the most difficult questions in the philosophy of mind. How can we be sure that other individuals share the same psychological framework with us? If I see a grimace of suffering on another person's face, can I be sure that she is suffering as I would if I had the same facial expression? If I look at something because of its color, is the feeling I get from the color purple the same as other people? These kinds of problems require complicated answers that often prove unsatisfying. But these are, after all, questions that affect other grown people. With young children, it is even more difficult. With other animals, they become almost exasperating. So what about the problem of other minds when such minds are artificial?

The chapter addresses these questions, suggesting that while most of the feelings we have about humanoid robots are inspired by fear or a desire to enslave them, the best approach would actually be one inspired by trust. Trust in robots, on the other hand, is a feeling that may have a natural basis. Anthropomorphism in particular – when properly understood – is a constructive sentiment that can ground the relationship between humans and robots. In the last part of the chapter, this kind of reasoning is tested on some very delicate cases, such as the relationship with weapons that can kill without human choice, and the sentimental and sexual relationship with sex robots. Perhaps in the future we will have to learn a different sentimental grammar than the one we are used to today, or we will have to harbor the same kind of distrust towards sex dolls that seems to prevail today towards weapons capable of killing without any feeling of compassion or even anger, as is typical of conflicts between humans.

6.1 Fear, Enslavement, and Trust

The relationship between humans and machines has always been a source of both opportunities and problems. As is well known from both classical studies (Leroi-Gourhan, 1964) and more recent investigations from cognitive archaeology (Mithen, 1996), the style of human cognition is deeply influenced by the use of tools. From an evolutionary point of view, machines can be considered as particularly advanced and efficient tools, or prostheses. And yet, as the tools have become highly ingenious machines, the usual feeling of satisfaction on using them is accompanied by one of apprehension. More generally, the most common reactions to autonomous machines and artificial intelligence are twofold: fear of machines and the desire to enslave them. On the other hand, one can be confused to realize that besides living beings there are also artificial creatures that can move autonomously and seemingly have their own goals and plans.

Our evolutionary history has not prepared us for this. Until the recent past, the tools used by individuals to achieve their goals did not have a human-like appearance. Objects such as knives, plows, and lenses are wonderful ways to enhance human capabilities, but they do not compete with or resemble humans. As we will see below, the main reasons that lead us to consider something as another mind are all biological in kind. We tend to consider something as a mind when it is endowed with faces, eyes, meaningful expressions and movements which are similar to those we see in humans and other animals. This spontaneous attitude is hard-wired into the human brain, because until a few decades ago, that is, a trifle in the deep time of evolution, it had never come across any artificial system that could have led us to think it was equipped with any form of real mind.

Besides curiosity, the most instinctive reaction to the possibility of machines with intentional performance is fear. The unpredictability of the behavior of a

Figure 6.1: The three automata built by Henri-Louis Jaquet-Droz. On the left *the musician* (1774), in the center *the draughtsman* (1774), on the right a view of the control mechanism of *the writer* (1772), a programmable handwriting automaton. (Musée d'art et d'histoire, Neuchâtel).

machine that seems to have purpose and to perform meaningful actions was the basis for the feeling of fear that is characteristic of the first phase of the relationship between humans and the machines they create. The automata produced in the golden centuries of the modern age were not only delightful toys for social entertainment, they were also a way of imitating living organisms by mechanical means, trying to resemble the Ruler of nature. In this spirit were built female doll organ player and an amazing draughtsman, both built by the Swiss watchmaker Jaquet Droz in the 18th century, and many other amazing mechanisms. Automata were a consequence of the philosophy of mechanism, which was also so popular in physics, especially classical mechanics. If the whole world, including living bodies, is nothing but a magnificent mechanism, the desire arises to produce small mechanical devices that serve as proof that even the human imagination is somehow connected to the universal mechanism that the very 'blind watchmaker' uses to shape the world. Watches, mills and the first machines of the Industrial Revolution were indeed the preferred models for imagining how not only the human mind, but also the body, works.

Even when, at the beginning of the 19th century, the notions of 'organism' and 'history' entered the scene of thought as hegemonic models for explaining the diversity of natural phenomena, including the functioning of language. People still believed that at the microscopic level, living organisms were nothing more than small machines, and that at the macroscopic level, the great human narrative progressed only as giant windmills driven by relentless and repetitive laws. The Frankenstein myth is a testimony to the prevailing sense of fear that animated the relationship between humans and their artificial creatures. Frankenstein can be considered as the champion of the feeling of fear towards machines; he is the archetype of human fear that machines can turn against their creators. Frankenstein is also a warning, a kind of memento mori of what can happen if you distort the biological organization and turn it into something completely artificial. It is no coincidence that this warning also underlies the first law of robotics of Isaac Asimov. In the science fiction story "First Law" (Asimov, 1956), the Three Laws of Robotics are as follows:

> 1: A robot may not injure a human being or, through inaction, allow a human being to come to harm; 2: A robot must obey the orders given it by human beings except where such orders would conflict with the First Law; 3: A robot must protect its own existence as long as such protection does not conflict with the First or Second Law; The Zeroth Law: A robot may not harm humanity or, by inaction, allow humanity to come to harm.

> (Asimov, 1956)

In a word, robots have to be mere slave for humans. Very etymological, one could say.

In Rossum's Universal Robots, the celebrated play by Karel Ĉapek, he used the term 'robot' for the first time by referring to humanoid androids made of synthetic organic matter. The basis was the Czech word 'robota', meaning 'forced labour' or 'slave' (Ĉapek, 1920). When robots are not endowed with human traits, they do not cause so much fear. Consider the extraordinary development of industrial automation, which is capable of transforming handicraft manufacturing into real mass production. It has provoked many negative reactions since the very beginning of the Industrial Revolution – e.g., the Luddism movement – but in the end, these are social reactions, not a genuine psychological response. Things change when robots become human-like creatures. Humanoid robots do indeed trigger feelings that industrial automation normally does not cause. People are sometimes frightened, as in many Hollywood movies.

The story usually goes as follows. The first step is characterized by success and satisfaction with the result. The machine performs some harmless tasks; in fact, these are usually entertaining and socially beneficial behaviors. In the second phase, the robots refine their performance without following a series of explicit instructions. They learn independently from social interactions and the environment. The final step is the revolution against humans. Once robots attain a form of consciousness, they become aware of their enslaved nature and begin to imagine their own purposes, which are generally different from those originally assigned to them. In general, the process of emancipation from human beings has a topical moment. The machine becomes so refined that it eventually looks at the world with the same eyes that humans look at it with. From this moment on, i.e., when the automata become aware of their own enslaved nature, the machine perceives Asimov's Laws as unacceptable and rejects them.

Singularity, the science fiction film from Robert Kouba (2017), contains in its title a reference to a mathematical idea behind which is the fear that our slaves may rise up against their masters. The basic idea is that if an artificial system becomes smarter than humans, it should also be better at designing new systems. The result should be a recursive loop toward ultra-intelligent systems (Good, 1965), with acceleration reminiscent of mathematical singularities (Vinge, 1993). From a cognitive point of view, singularity is like an earthquake. According to David Chalmers, if "there is a singularity, it will be one of the most important events in the history of the planet. An intelligence explosion has enormous potential benefits: a cure for all known diseases, an end to poverty, extraordinary scientific advances, and much more. It also has enormous potential dangers: an end to the human race, an arms race of warring machines, the power to destroy the planet" (Chalmers, 2010, p.9). One can be skeptical about this enthusiasm. Perhaps, as the design of artificial cognitive architectures goes on to address more demanding challenges such as artificial consciousness, we are heading for a slowdown effect instead of the expected acceleration effect (Plebe and Perconti, 2013). Things could even get worse, i.e., towards a scenario where intelligence, instead of surpassing human intelligence, and instead of slowing down in its development, could end up 'breaking down' (Plebe and Perconti, 2020).

There is an interesting difference in the reactions that humanoid robots trigger in Western and Eastern cultures. While robotics is equally developed worldwide, humanoid robotics is characterized by a strong Japanese supremacy and Eastern robotics in general. In Japan, relations with humanoid robots are generally well respected. Projects aiming at building humanoid robots capable of caring for other people, especially the elderly and sick, are strongly encouraged. The reason for this different social consideration of humanoid robots in East and West can be questioned. In the absence of extensive research on the subject, one could perhaps only make some speculations inspired by a variety of cultural stereotypes. It can be assumed that the well-known preference given by the Japanese to impersonal relationships, with the typical massive use of vending machines and any electronic devices bypassing emotionally hot human relationships, plus the typical Japanese discretion towards the most intimate and personal sphere, results in the elderly and sick people preferring to deal with machines in the more embarrassing aspects of care relationships. But, of course, this is just speculation, perhaps overly conditioned by the most common cultural stereotypes.

Swinging between the feeling of fear and the will for enslavement, a third alternative seems to be more promising. For a harmonious relationship with humanoid robots, we need to cultivate the natural attitude of trust. Trust, in fact, is a key feature of social cognition, both when it concerns other people and when it involves other animals or machines. The point is that, to have a productive relationship with a given system, we must treat it as a rational agent. It does not matter, after all, that it is really this way, or not. Take the case of Shakey, a robot developed at Stanford in the late 1960s (Figure 6.2). Shakey is the archaeology of artificial intelligence. It was a fairly simple robot, consisting of a camera mounted on a wheeled device and connected to a computer via radio. The robot was able to distinguish boxes from other objects in space, such as pyramids, by converting the images captured by the camera into sequences of 0s and 1s. Shakey had no internal representation of a box prototype in a photo archive so that he could compare the prototype to the actual box. He was only able to track the vertices in the drawing he had obtained with pixels 0 and 1 and match them to figures such as the letters L, T, or Y. If there was a Y, Shakey was able to 'understand' that it should be a box because the image of a pyramid cannot produce a Y vertex. Shakey was not a rational agent, but he apparently could be.

But is the gap between looking like a rational agent and being one so important, after all? According to Dennett's intentional stance, no. Intentional stance simply means treating a system whose behavior one is interested in predicting as if it were a rational agent endowed with intentional states (Dennett, 1987, p.30). Such an attitude is not the only resource we have for predicting the behavior of a given system. According to Dennett (1987), we can use two other styles of thinking for this purpose: the physical stance and the design stance. We use the

Figure 6.2: The robot Shakey.

physical stance when we know the physical nature of an object and can thus draw on the predictive capabilities of the hard sciences, such as physics and chemistry. For example, if we force a rubber ball under water, we can predict that – once released – it will quickly float back to the surface. In other cases, it is better to take the design stance. In this case, it is not important to know the physical nature of the device in order to have a productive relationship with it. It is enough to know how it is designed. This attitude is best suited to artificial systems such as cars, microwave ovens, or printers. No intentional state needs to be ascribed to them.

It is quite different for persons. By far the most convenient way to predict people's behavior is to ascribe intentional states to them, that is, to treat them as if they were rational agents. This method is not only the most convenient, it is the one we rely on all the time. We all surrender our lives to intentional stance every single day. Let us say you are crossing a crosswalk when a car approaches. Normally, in such a case, we cross the street despite the approaching car because we assume that the driver has seen that we are at the crosswalk and we believe that he or she has a belief that we should let   pedestrians pass when they cross the street at the crosswalk. We trust a belief that we have attributed to a person we may barely see, and we are confident that this belief plays a causal role in his or her behavior. Trust is everything in the case of an intentional stance. Intentional stance does not involve any ontological commitment to the system we are dealing with. It does not matter, in other words, that the system is really a rational agent, endowed with the most basic elements of folk psychology, such as beliefs, desires, and intentions. All that matters in adopting the intentional stance is the advantage that one acquires thanks to it, about one's predictive abilities towards other people's behavior and the possibility of making good behavioral predictions. Although Dennett is initially trying to describe a stance we naturally take when we are interested in interpreting human behavior, the same stance is at stake when we try to account for any rational agent. In this way, Dennett is a naturalized follower of Turing's 'imitation game', as we saw in §2.3.

Trust is a key component of the intentional stance. It is the attitude of attributing to the rational agent behaviors that are sensible and proportionate to your situation. If you call a restaurant for home service, you have no reason to believe that the promise to bring you food is a joke. Why, after all, should the restaurant manager kid us? It would go against her interests and compromise the reputation of her business. Trusting the food (sooner or later) to be delivered to us is a completely reasonable attitude and the only one that makes the relationship between the customer and the restaurant meaningful. Trust plays the same role in social knowledge as the principle of charity plays in communication intercourses. When we talk to people, what we are talking about often has many possible interpretations. But only one is usually adopted as the one that corresponds to the intention of the speaker. Why that one? The answer depends on the principle of charity, which requires us to choose – among the various possible interpretations – the one that best suits the rational framework that we are committed to assign to our interlocutors. We are obliged to be charitable towards our interlocutors because this is a condition making an effective communication possible. From the pragmatic point of view, being charitable and trusting other people are two conditions of possibility of human communication and social cognition, respectively.

Social cognition is the most typical feature of the way humans acquire and process their knowledge about other individuals. We are not the only social animals. Many insects, like ants, and many mammals, like beavers, are also social animals. Nor we are the only ones who use knowledge to shape our sociality. On their hunting trips, wolves, for example, use the knowledge they have about conspecifics, prey and territory to increase their chances of success. The form that social cognition takes in humans is, however, something specific. It is strongly based on language and, thanks to it, has outstanding performances. For example, the way human signals are context-dependent allows them a flexibility that is unknown in the animal kingdom. The way we use context-dependent signals, such as indexicals and pointing gestures, gives human social cognition a specific character (Perconti, 2002).

What about social cognition in human-machine interaction? How does social cognition work between human beings and the machines that we build? The short answer to these questions is: more or less in the same way that social cognition works among humans. In fact, we do not have different ways for social cognition to be adapted according to the different needs of the circumstances. There is no one social cognition for males, one for females, one for children, one for animals, one for pets, and one for machines. The only way we are able to use social cognition is through the ability to interpret the behavior of others through the intentional vocabulary. We make sense of the behavior of others, as well as our own, by using the vocabulary of folk psychology, made up of 'beliefs', 'desires', and 'intentions'. We use mentalization, simulating what is likely to pass through the mind of other people, in order to make appropriate behavioral predictions based on what we ourselves would do in counterfactual circumstances.

Social cognition happens in two ways. On the one hand, it works automatically, quickly, and without any consciousness. On the other hand, it works in a deliberative, conscious, and slow way. The first way of running regulates the sensory-motor interactions, deriving the sense and goal of the actions from the way they are performed. If someone puts the needle of a syringe into a hand, we can almost feel the same pain they are probably feeling (Avenanti et al., 2005). To experience this sense of empathy, no conscious deliberation is required. It only takes the same kind of neural networks in our brain that are activated in the individual we are observing. Fortunately, this happens automatically, as studies of canonical neurons and mirror neurons have shown, as well as the observation of behaviors that reflect the ability to experience emotional contagion, such as when we yawn or laugh simply because the people around us are doing the same.

The body is full of many 'motor triggers', like the patellar reflex. When the knee is hit with a small hammer on the tendon below the kneecap, the leg is pulled up suddenly and without any deliberation. For the intentional processes to be activated, things work in a similar way. There are some 'mental triggers' that elicit the intentional interpretation of the observable behavior. The 'mental triggers' are those natural mechanisms that are automatically activated by certain classes of environmental stimuli and that produce, as a result, the use of an intentional interpretation in understanding behavior (Bruni et al., 2018). The group of mental triggers includes the predisposition to classify in a different way the living creatures from those that are not on the basis of the kind of movement they perform, the disposition to recognize in certain perceptual patterns a given meaningful face, and the ability to share the attention with others. There are more mental triggers, but these are probably among the most significant in those which we may come across. If we come across something that stimulates our brain with the just mentioned three mental triggers, then we are prepared to engage in an intersubjective mode and are ready to activate the social cognition processes.

6.2 Artificial Anthropomorphism

What we have described until now is a kind of anthropomorphism. In particular, it is the way anthropomorphism appears as a natural attitude able to elicit social cognition. Anthropomorphism is not like many other '-isms' one can find in the market of theories. Normally, when faced with a given '-ism', it is a matter of choosing whether to grant it our favor or reject it. What about skepticism, feminism, or anti-speciesism? It depends on the set of beliefs each individual holds, on their consistency, as well as on the emotional and social conditioning to which each individual is subjected. With anthropomorphism things are different. The tendency to give a human form to the systems we happen to deal with in our daily routine is not, in fact, a social construction. Rather, it is something that is found, from the very beginning, within us. Arbilly and Lotem (2017) argue for a similar claim with their *constructive anthropomorphism*. In their words, "We believe that the natural tendency of using our human experiences when thinking about animals (i.e., the tendency to anthropomorphize) can actually be harnessed productively to generate hypotheses regarding cognitive mechanisms and their evolution" (p.2). Anthropomorphism is a hallmark of our mind. It is, indeed, a condition that allows social cognition to take place. It is like trust and the principle of charity in interpersonal communication. These are natural trends, with a biological basis in brain working that have been better understood in recent decades, which are responsible for the right functioning of social cognition.

Broadly speaking, if we are interested in humanizing technology, anthropomorphism should be seen as a natural way to make sense of other people's behavior through the use of intentional vocabulary (Bruni et al., 2018). Indeed, to foster an ecological exchange between humans and robots, the designer must consider what can truly enable a natural human-robot relationship. And the best candidates for this role are again things like the ability to share the attention of other humans, follow their gaze, express meaning in their faces, and move similarly to other living creatures. To this end, the anthropomorphic mindset could inspire computer scientists as they are busy trying to find the right computational architecture that will allow a humanoid robot to have fruitful interactions with a real human. This ecological concern should inspire attempts to humanize both the body and mind of robots. Sandini and Sciutti (2018, p.2) emphasize the difference between 'illusory humanizing robots' and the challenge of making them more 'humane': "A humane robot is a robot that is considerate of humans, i.e., a robot that maintains a model of humans to understand and predict human needs, intentions, and limitations while being transparent, legible, and predictable".

It is interesting to note that the attributive mechanisms mentioned above, i.e., anthropomorphic mental triggers, are good guides for designing both humanoid bodies equipped with the right mentalization cues and human robot heads equipped with the same cognitive abilities to detect these cues in overt human behavior. Virtual agents can mimic human eye abilities with greater accuracy than embodied robots (Ruhland et al., 2015). Things look a little worse for humanoid robots. Even the best robots are not yet capable of pupil dilation, although this behavior is an indicator of the mental state of people (Hyönä et al., 1995; Hanson and White, 2004; Admoni and Scassellati, 2017; Chevalier et al., 2020). Among the (anthropomorphic) mental triggers is the ability to detect the movements performed by biological agents in the environment (Johansson, 1973). It seems that human infants of less of 3 months of age are endowed with this capacity (Simion et al., 2008). The same ability should be given to humanoid robots if we are interested in an ecological Human-Robot Interaction (Vignolo et al.,

Figure 6.3: The iCub humanoid robot (Italian Institute of Technology and Plymouth University).

2017). The idea is to equip robots with a module integrated in the software as, for example, in the iCub humanoid robot software (Vignolo et al., 2016). The iCub robot is designed as a baby humanoid, as visible in Figure 6.3, and is aimed at developmental robotics research (Cangelosi and Schlesinger, 2015).

The general point here is the possibility of considering anthropomorphism as the natural attitude of ascribing human psychological properties to other systems in order to make sense of their behavior. The main advantage of adopting this stance is the use of a fast and frugal heuristic (Gigerenzer, 2007; Gigerenzer et al., 2011). One does not need any complicated calculus capable of summarizing all the environmental cues in the right way; it is enough to try to attribute the key components of human nature to the system we are interested in, to predict her or his behavior, and see what happens. Anthropomorphism is an important method that the human brain uses to make tractable neural computation. Computationalism claims that mental processes are computations (Piccinini, 2009), that is, that a cognitive process expresses one or more computable functions (Cummins, 2000). This means that computations must be functions that are computable using an effective procedure or algorithm (Perconti and Zeppi, 2014). However, even if not every function is Turing-computable and there are more or less precise bounds on what we can call computable, this does not seem to be a real limit.

Computability tests, for example, are inherently unbounded, and this is because they rely on an idealized computational model, or a Turing Machine, that can de facto draw on infinite time and space resources (an infinite tape and an infinite computational loop, in archeological terms). A function is computable in the classical sense, if there exists a finite effective procedure, i.e., an algorithm, that stops after a finite number of steps, consumes a finite amount of memory, and works for an arbitrarily large set of inputs (Enderton, 1977). Such a solution must somehow be finite, but there is no a priori indication of how much memory or time it requires to provide an output.

There are significant tensions here. On the one hand, traditional Turing machines require at least virtually infinite space-time resources. On the other hand, neither humans nor their machines have infinite resources at their disposal. Human brains must therefore address the problem of computational tractability. The same must somehow be done by the artificial machines we use to perform cognitive tasks. Neurocomputation has the same limitations and requirements as artificial computation. For this reason, the human brain uses several frugal heuristics to deal with the problems it experiences in its environment (Zeppi and Blokpoel, 2017). Among the most important frugal heuristics from a computational point of view are the mental triggers mentioned above. They are like bets that the brain makes about the world. If something moves like a living organism, if we can observe something in the form of a face in which we can detect emotions and feelings, if it is possible to share attention with that system, then we can bet that we are dealing with a rational agent. Sometimes we will be wrong. But most of the time, the bet is won. And in the gain that this difference makes, in terms of time and from a metabolic perspective, is why it is worth the bet. Anthropomorphism is that kind of bet. Sometimes it causes us to be a little too naive and see the world through cartoonish eyes, but most of the time it wins us time and cognitive energy.

6.3 Robots and Sex

Artificial anthropomorphism affects all major areas of human-computer interaction. When interpreted in the constructive manner mentioned above, it can become a constraint on design and play an important guiding role in trying to make the relationship with humanoid robots more ecological. One elective field to appreciate the ability to ecologize the relationship between humans and humanoid robots is that of sex dolls. In the twentieth century, these were mostly comical inflatable dolls with extremely accentuated feminine features, albeit in a very naive way. Scenes of degradation and wimpy or perverted people come to mind. Today, things have changed radically. We are faced with sophisticated robots capable of evoking more complex emotions than the primitive feelings characteristic of the very first sex dolls. Take the case of the products of RealDoll, one of the most competitive companies on the market. Their Harmony, priced at several thousand dollars, promises the buyer an experience closest to the real world. Similar performance, backed up by artificial intelligence, is promised for the Samantha robot, made by Synthea Amatus Doll, shown in Fig. 6.4. Sex robots are now so close to reality that they raise challenging philosophical and social questions. Some people fall in love with their 'toys', or at least claim to be infatuated with them. Others choose them as 'partners' in their lives instead of humans. Some intellectuals have begun to view them as political subjects to be protected, endowed with rights that would be trampled by the predatory mentality of capitalist society.

Figure 6.4: Samantha, by Synthea Amatus Doll.

The Campaign to Stop Killer Robots was launched in 2013 and aims to ban lethal autonomous weapon systems, artificial systems that can kill humans without the explicit intent of others. Such weapons include drones, missile systems and autonomous robots, and can be designed to be defensive or offensive. The most ethically controversial point, of course, is giving an artificial system the 'choice' to kill a human individual. But it is the most attractive point from the soldier's point of view—-to eliminate the responsibility of killing other humans. To be honest, it is not a real choice for the robot, but merely the activation of a certain sequence of behavior on condition that a given set of parameters is met. From the human perspective, however, the scene appears completely analogous to a decision by an artificial system to kill one or more humans. It is precisely the detachment of 'choice' from human will and responsibility that has led many countries, as well as several non-governmental organizations and over 1,000 scientists, to call for a ban on such technologies, thereby preserving human control over the use of military force. Once the taboo of the 'artificial choice to kill' is overcome, the path to weapons without moral responsibility is permanently followed by all those who wish to simultaneously free themselves from moral responsibility and gain efficiency in war scenarios.

Following the example of the Campaign to Stop Killer Robots, another campaign against artificial autonomous systems was launched in 2015: the Campaign Against Sex Robots. This time the target is a seemingly harmless kind of robot, namely sex robots, or sexbots, as they are sometimes called. If war with robots is not appropriate, at least you can make love, you might say. But Kathleen Richardson, an expert on the moral implications of robotics, has decided to lead the Stop Sex Robots campaign to prevent that possibility as well. Her point is that relationships with robots are not ethically neutral. The fact that robots are conceived as objects and categorized in terms of their attributes, and yet have human traits, is indicative of the implicit consideration of human relationships. The point is that we are morally responsible for this implicit consideration.

According to Richardson, love and sex are relationships that can only occur between people. To apply these categories to what happens between a human and a robot is a categorical mistake. The proper category to describe this type of relationship is exploitation (Richardson, 2015, 2018). To ban sex robots is to condemn exploitative human relationships as well. According to the Montreal Declaration for Responsible Development of Artificial Intelligence, "technologies should not encourage cruel behavior towards robots that take on appearance of human beings or animals and act in a similar fashion" (Coghlan et al., 2019). Richardson's argument, however, is plagued by several problems. First, it is not clear how to conceptualize the relationship between what can be done with a robot and what is considered acceptable in the context of human relationships. Consider how we use scarecrows. They are humanoid puppets, much like the farmers who build them, and are left to the mercy of weather and animals in the fields. If we were to adopt Richardson's line of reasoning, we would have to ask, "What perverse model of human relations does the construction and use of scarecrows reflect?" Should they not be banned to prevent the spread of a model of human relations that seems clearly based on the exploitation of other humans?

Moreover, the ethical link between fiction and reality that Richardson proposes can also form the basis for an argument that leads, however, to opposite conclusions. The point is to use the argument of catharsis, typical of ancient Greek theater. The most socially dangerous human impulses are portrayed on stage in such a way that the same tension is resolved in the real world. Thus, although many stories of violence and oppression are performed, theater is able to reduce the level of violence in society. Similarly, sex robots can be seen as tools for creating a fictional erotic scene in which the subject actually acts, rather than doing so in the real world. In this way, sex robots could replace prostitution and play an ethically positive social role (at least considering the aspects of oppression often associated with this practice).

However, even if we abandon the idea that there is only a symbolic connection between sex robots and the way we conceive of human relationships, the problems are far from solved. How should we view the desire for some form of social and legal recognition for the relationships established between humans and robots? We do not yet have a clear idea of how to consider the rights of other species. It seems far too early to come up with an agreed theoretical framework that might govern, within common sense, the sentimental relations between humans and their robots. Is this perhaps a paroxysmal form of fetishism? If fetishism is ultimately nothing more than the fixation of the sexual drive normally reserved for other humans on an object, then robots can be seen as the ultimate landing ground of fetishism. If fetishistic interest in shoes, for example, follows the path of metonymy, in which the part stands for the whole, love of robots reclaims the whole. It does so, however, in an entirely fetishistic dimension in which everything is symbolized and ordered in order to achieve a set of goals that are set before the relationship that it must try to fulfill. After all, Pygmalion also fell in love with one of his sculptures. The experience of falling in love with a doll is much older than one might think (Lee, 2017).

Chapter 7

Towards the Hardest Things

The possibility of machines gaining consciousness of themselves is among the most intriguing challenges in cognitive science. On the other hand, it is also one of the most disturbing scenarios for those who fear the autonomy of machines. Indeed, when you buy into some form of machine consciousness, you also end up buying into strange ideas, such as artificial morality and the need for legal rules defining what is and is not allowed when dealing with a kind of machine that eventually looks more and more like ourselves. This chapter is devoted to just such new perspectives and outlines their difficulties, both theoretical and social. It considers several areas of application, including that of the algorithms that control the autonomous driving of next-generation vehicles.

The idea that machines are endowed with a an inner life of their own, that they feel something in who they are, that they engage in terms of individual responsibility, that they are – in short – similar to humans, sounds like the ultimate landing pad of the artificial intelligence adventure. All of this still seems futuristic in some ways, and yet that is exactly what is happening now. Preparing for such scenarios is a social responsibility for cognitive science and artificial intelligence, at least if we do not want to face unprepared a technological development that in some ways already seems to be taking place without any explicit human control.

7.1 Self-consciousness and Social Cognition

According to the common sense view, self-consciousness is the climax of human cognition. If we could save only one function of our minds, among attention, language, touch, memory, or self-consciousness, we would have no doubt. We would want to remain self-conscious creatures and give up the other faculties. Self-consciousness, indeed, provides the feeling of being something special in the animal kingdom and the rest of the world. Moreover, self-consciousness is the seat of personal identity for each of us. If we were to lose our sight or our memory, we would still be one of those who suffered a loss, but if we remained without consciousness, every other gift of the mind would no longer concern us. Self-consciousness is the basis upon which every other faculty of the mind acquires personal value. However, we doubt how much nature cares about such a satisfying feeling of being self-conscious.

Nature, of course, cultivates no interests. Nature simply has no mind and it is not a person. Therefore, it has no interests whatsoever and cannot care about anything. The impression that nature might have some sort of interests comes from the fact that we observe changes in it. Since it appears that the changes are directed in a certain direction, we are inclined to think that they are the result of a design, that is, of a mind. If there is a plan, there will be a mind — so goes the tenet of the advocates of intelligent design. Natural change, however, is merely the result of the mechanism of natural and sexual selection. It is Charles Darwin's wonderful mechanism of evolution that is solely responsible for the changes we observe in nature.

If nature has no interests, then it has not given us any sense of being special in the world, as self-conscious creatures. Self-consciousness must therefore have some other purpose. Let us now try to formulate this argument in the form of a question. What is the purpose of self-consciousness (if any)? Could not our brain perform its functions just as effectively, but without the sensation of consciousness? Several considerations lead to a positive answer to this last question. First, consciousness and self-consciousness are extremely expensive cognitive functions. As Benjamin Libet's (2004) seminal studies have shown, becoming aware of a particular cognitive process is something that takes a very long time. It seems that before the brain encodes a particular motor sequence, it prepares it in the background. This is the so-called readiness potential (from German, *Bereitschaftspotential*) (Kornhuber and Deecke, 1965). Libet found that for simple motor sequences requiring a deliberative choice (and thus free will), the readiness potential precedes the feeling of a free decision by about half a second, that is, a very long time in terms of the nervous system.

It is as if the automatic brain, which is much faster, is preparing offline, the most likely scenarios for the advent of the much slower deliberative and conscious brain. In some cases, this delay extends to as much as eight seconds (Soon et al., 2008, 2013). Time is an important resource for the brain, and all the effort to manage the time it takes to awareness is terribly expensive. On the other hand, self-awareness is just as expensive metabolically, i.e., in terms of the blood flow and glucose the brain needs to function efficiently. There doesn't seem to be a center for consciousness in the brain, but we know that being conscious costs a lot of energy. The brain counts for only 2% of the human body's mass, but it demands about 20% of the body's total energy (Magistretti and Allaman, 2016). Consciousness requires a significant part of this energy budget (Pepperell, 2018). Having consciousness costs metabolism and time. It must be worth it, you might say.

But there are many examples of cognitive tasks being performed efficiently, yet unconsciously. Think of how much visual information is processed as we walk down the street without being aware of anything. We can mind our own business, free to wander with our thoughts while the brain, without telling us, considers a thousand obstacles that the terrain puts in our way. The best examples, however, come from the field of neuropsychology. The cases of 'blindsight' and neglect show how the brain can correctly process a lot of information unconsciously in order to for us to adapt to the environment, and that this is information we think we are consciously aware of. We may even swear that we want to live in one house and not another on a mere whim, when in fact our brain has chosen that particular house because the other is on fire, as depicted in Figure 7.1 (Marshall and Halligan, 1988). In short, our brains allow us to be efficient, even without self-consciousness, in performing tasks in which we are confident that it is consciousness that is guiding us. Could we not always work this way?

The reason why, both from an evolutionary perspective and from the point of view of cognitive architectures, we are conscious creatures is that self-consciousness serves the purposes of social cognition. The ability to regulate social relationships by attributing mental states to other people and interpreting their behavior by means of intentional vocabulary is an evolutionary advantage of crucial importance to humans. For this to work properly, however, there needs to be an internal logical space in which to test and construct possible scenarios

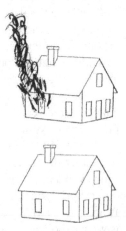

Figure 7.1: Although the patient had awareness only of the right side of the house, her brain unconsciously took into account that in the left side of the image the house was on fire.

off-line. Such an internal logical space is what we commonly call 'selfconsciousness'. In other words, self-consciousness is the logical space that social cognition needs to do its job.

This hypothesis can be called 'mindreading priority account', stating that mindreading evolved prior to self-consciousness and that the latter is for mindreading purposes (Perconti, 2020). According to Peter Carruthers et al. (2012, p.16), 'Self-knowledge results from turning our evolved mindreading capacities on ourselves'. One might ask for what. We need a reason to suppose that turning mindreading on ourselves was something worthy for adaptation. Mindreading priority account claims that the reason why nature pays a cost for self-consciousness lies in the needs of a logical space which allows mindreading to work off line and make simulations. To take place, high-level simulation must have an inner space as a base from which it operates. Self-consciousness is the inner space from which high level simulation proceeds in its behavioral predictions, and in understanding the reasons behind the actions of others as well as our own.

In arguing for the mindreading priority account, all depends on what you mean by the expressions 'mindreading' and 'self-consciousness'. And, it also depends on what you mean by the word 'priority'. According to James Daw:

> Mindreading priority could be: a conceptual priority-mindreading can be conceived independently of self- consciousness; a functional priority-mindreading can be posited as a cognitive mechanism independently of self-consciousness; a developmental priority-mindreading develops in infancy prior to self- consciousness; a neuroscientific priority-mindreading is a brain function that is distinct from and active temporally prior to self-consciousness; an evolutionary priority-mindreading evolved prior to self-consciousness and self-consciousness is a byproduct of mindreading.
>
> (Dow, 2012, p.42-43)

The problem of the priority between self-consciousness and mind-reading is not, however, a genuinely conceptual affair. Indeed, from the standpoint of conceptual analysis, there is nothing to prevent us from thinking that social cognition is prioritized over selfconsciousness, but also its opposite. Rather, this is an empirical matter. It is first and foremost a matter of seeing how things are in the course of evolution, both from a phylogenetic and ontogenetic perspective. In child development, it is important to distinguish between low-level mentalization, which is characterized by behaviors such as emotional contagion and motor empathy, and high-level mentalization, which is the ability to imagine what it is like to be someone else and what behavior we would adopt if we were in the shoes of the person we are observing. This second level of mentalization is based on skills such as pretending or using counterfactuals. Similarly, it is important also to distinguish between two levels of self-consciousness, namely self-recognition and reflective reasoning. If we keep these kinds of distinctions in mind, we can give a non-simplifying answer to the question of the priority between mindreading and self-consciousness in child development. Low-level social cognition develops as early as the first few months of life, when the child is already capable of empathy and social   contagion, and also has the ability to make behavioral predictions about the movement of other people. In the second year of life, children gradually become able to understand that others do not necessarily know what she knows or perceives. These kinds of reflections cause the child to acquire the capacity to experience herself as an autonomous subject in relation to the world. The child thus becomes self-conscious. For example, at around 18 months, she is usually able to recognize her own image reflected in a mirror.

To bring up the issue of phylogenetic priority as well, it is remarkable that this mirror recognition ability has been detected in the 4 species of apes, as well as in magpies, dolphins, and partially in elephants. It appears to be an uncommon ability in the animal kingdom and is reserved for animal species that are quite similar to humans and have sophisticated forms of social cognition. Once an initial capacity for social cognition and self-consciousness has been developed, the child is ready for the more sophisticated forms of this capacity. Thus, the different components of social cognition and self-consciousness, such as counterfactual simulation and stream-of-consciousness, go from month to month until their interweaving eventually constitutes the characteristic form that interiority has in the human mind.

Mindreading priority account is based on two theoretical tenets: intentionalism and dual-process account. Intentionalism is the thesis that the mind is by and large made up of representations and that intentionality is the key feature of mental representations. Cognitive processes are information processing whose bearers are precise mental representations. Intentionality is a subpersonal property. It is the property that representations should have, that they refer to an internal content presented according to a mode or format of representation. This perspective has its modern roots in John Locke's theory of ideas, but it reached its standard formulation in 19th century theories of mental representation. At the beginning of the century there was the first complete formulation, thanks to philosophers such as Kant, Reinhold and the theorists of Kantian grammar (Perconti, 1999). The second famous formulation, at the end of the century, was due to Brentano. In the middle is the whole psychologization of the theory of mental representation in a reliable basis for the incipient empirical psychology.

The second theoretical preference is for the dual-process account, which is based on the idea that there are two distinct processing modes in the brain, namely System 1 and System 2 processes (Evans and Frankish, 2009; Evans and Stanovich, 2013). System 1 includes mental processes that are unconscious, fast, and automatic. System 2 involves conscious, slow, and deliberative mental processes.

> What dual-process theories have in common is the idea that there are two different modes of processing, for which I use the most neutral terms available in the literature, System 1 and System 2 processes.
>
> (Evans, 2008, p.256)

Following this line of reasoning, the concept of self-consciousness should be divided into two components, as was advanced earlier. We do not normally experience self-consciousness as something that exists in degrees or has components. The experience of being self-aware tends to be something immediate and unarticulated. It looks like an 'all-or-nothing' phenomenon: if someone is self-aware, they are fully so; otherwise, the subject is not conscious at all. The personal effect of self-consciousness is, in a sense, comparable to turning on the light in a room at night. When the light is on, everything is clearly visible; when it is off, everything in the room is plunged into darkness.

Similarly, when we are self-conscious, all the other aspects of mental life – moral and legal responsibility, the ability to calculate, and much more – can take place. But when our self-consciousness disappears, as a result of sleep or death, we are no longer responsible for anything, nor can we exercise any form of thought or deliberation. We are no longer alive and whatever happens around us is no longer our concern. We must, therefore, make distinctions that can better analyze the ordinary notion of self-consciousness. To do this, we might consider, on the one hand, all those faculties which have something to do with self-identification, and, on the other, the various kinds of streams of consciousness.

Self-recognition is the simplest form of self-consciousness, and, from an evolutionary point of view, also the oldest. It presupposes that the subject is able to refer to himself by means of a reflexive representation, typically a schema of his own body. Reflexive reasoning consists of the ability to consider the role that a particular concept has within the reflexive network in which it is placed. Generally speaking, reasoning means being able to draw the proper consequences that follow from a particular thought we entertain. It is not simply having an idea, but recognizing that holding that idea implies some kind of responsibility to a set of other ideas. If the conversation is authentic, each of the interlocutors can appeal to the responsibility that the other has to the beliefs that they hold to be true. Usually this kind of appeal is public because it takes place in the social practice of conversation. However, the responsibility we have to the network of concepts in which we envelop ourselves also exerts its influence when we engage in inner speech, that is, in the private surrogate of conversation.

High level simulations use the same cognitive resources, such as counterfactual imagination, that are at play in reflexive reasoning. When we are engaged in inner speech, drifting silently in our stream of consciousness, we think in the same way as when we simulate another individual to predict his actions. High level simulation is an activity of projection that, in order to take place, must have an inner space upon which it can be based offline. In terms of self-recognition we must first consider the difference between body schema and body image. Whereas replication of body schema skills is essentially about matching some self-experienced sensorimotor features and the equivalent observed in an image, replication of body image skills requires social skills. Body image implies social recognition. However, this raises the problem of modeling social knowledge and background knowledge, with the familiar puzzles of the frame problem and the difficulties of modeling common sense knowledge. Taking into account, then, reflexive reasoning, the main problem is to model inner speech, which is the logical basis on which the conceptual side of self-consciousness operates. Inner speech consists of a temporal sequence of representations in which the attribution of a certain property to a reflexive representation has some consequences for the development of reasoning, both in terms of inferences drawn between thoughts and in terms of the experience of a new emotional 'tone'.

Considering reasoning in terms of inferences between concepts is an idea emphasized by Robert Brandom (2000, 2019). According to him, if we want to understand what it means to have a certain mental content, we should give much less importance to the classical notion of 'representation' and instead give a key role to the notion of 'inference'. The practice of asking for and providing justifications for our ideas constantly prompts us to establish the patterns of conceptual inference. This is precisely why we are responsible for our ideas, and when we are asked to provide reasons for what we believe, we articulate our thoughts inferentially. Indeed, reasoning can have ourselves as its object, in addition to other people and the events of the world (Davidson, 2001). I can become the object of my thinking and, in this way, experience my inner life. When we attribute a particular property to ourselves, we find that we thereby become responsible for other properties that are also relevant to us but which we had not yet considered. It is the web of the language system, with its rules and boundaries, that provides a framework for reflexive thinking and helps to expand the knowledge I have about myself.

This, of course, is not the only account of the development of the capacity for self-knowledge. According to Peter Carruthers, there are two general accounts based on the first and third person:

> One can then envisage two broad accounts of the evolution of a capacity for self-knowledge. One is first-person based. It is that self-knowledge evolved for purposes of metacognitive monitoring and control. On this account, organisms evolve a capacity for self-knowledge in order better to manage and control their own mental lives. By being aware of some of their mental states and processes, organisms can become more efficient and reliable cognizers, and can make better and more adaptive decisions as a result.
>
> (Carruthers et al., 2012, p.14)

The contrasting account is based on the third person and (p.15) "It maintains that the adaptation underlying the capacity for knowledge of one's own mental states is a mindreading faculty (consisting of a body of core knowledge about the mind or a domain-specific learning mechanism with representational primitives, or both), which evolved initially for social purposes."

7.2 Machine Consciousness

"Can computers become conscious or self-conscious?" is one of the most intriguing questions in current cognitive science. In answering this question, we can distinguish between ideological and pragmatic responses. In turn, ideological answers can be classified into positive ones ("Yes, it must be possible to build a self-aware machine") and negative ones ("No matter what happens in reality from the technological point of view, it simply must be impossible for a machine to become self-conscious"). These kinds of replies do not really meet the challenge contained in the question. It is believed the question is somehow superfluous, either because the answer can only be positive or because of the opposite. Those who support the negative answer in an ideological way include the 'mysterians'. Thomas Huxley was one of the old mysterians. His skepticism about the possibility of a scientific explanation of consciousness was based on the classical dualism between mind and body. The new mysterians, by contrast, no longer believe in the dualism of substances. Nevertheless, they remain skeptical about the possibility of science being able to provide a convincing explanation of what consciousness is. Colin McGinn (1991) is a typical new mysterian. His objection is radical and is based on anti-naturalization. In the words of Dan Hutto:

> His reasoning is that, as consciousness is essentially "non-natural" in character, there can be no real possibility of a scientific account of it in traditional terms.
>
> (Hutto, 1998, p.332)

This kind of skepticism is particularly entrenched in New York. The philosophy department at New York University, thanks to the work of Ned Block, David Chalmers and Thomas Nagel, has become the center of qualia-based objections to the possibility of providing a naturalistic explanation for the phenomenon of consciousness. Sheltered by the tranquility of Cambridge, Massachusetts, the Boston suburb where he works, Daniel Dennett ironically refers to the skeptical orientation of chaotic Manhattan:

> What support does McGinn offer for his striking conclusion that "there is something terminal about our perplexity" (p.7)? He draws his main inspiration from two philosophical sources, Thomas Nagel (formerly at Princeton, now at NYU) and Jerry Fodor, now his colleague at Rutgers. All three live in Manhattan and are no strangers to the Port Authority Bus Terminal. Perhaps this helps to explain their shared pessimism, for it does appear that we are witnessing the birth of a new school of philosophy: New Jersey Nihilism. Nagel's naysaying is renowned. His 1974 paper, *What is it Like to be a Bat?* advanced

the claim that it is impossible for us to know the answer to the title question. McGinn takes himself to be building on Nagel's foundations, expanding his claims, hardening the line. Fodor's nihilism is less often recognized, but has been emerging with increasing vigor in recent years.

(Dennett, 1991)

Port Authority Bus Terminal is an ugly building – unless you like brutalism and International Style – plunged into the heart of Midtown Manhattan. It has nothing to do, of course, with the New York philosophers' theoretical pessimism about the naturalization of consciousness. But it can be seen as a symbol, according to Dennett, of an ugly, pessimistic, almost 'terminal' theoretical trend. Among the ideological reactions to the possibility of building an artificial system that is endowed with consciousness are included the theoretical attitudes listed by Antonio Chella and Riccardo Manzotti (2020). According to Chella and Manzotti, the most common conceptual drawbacks in replying to the above question in a constructive way are computationalism, biological chauvinism, neural chauvinism, cognitivism, and bodyism. Computationalism, "the tendency to deal with either information or computation as they were additional substance that is somewhat generated by a computing machine", is a circular argument.

> In fact, we can use computations to describe any physical system (a star, a waterfall, a swarm of bees, a processor). The fact that we describe that physical system in computational terms does not change a physical system.
>
> (Manzotti and Chella, 2020, p.188)

Supporters of biological and neural chauvinism claim that consciousness is a property which emerges from a living substrate and therefore rule out the possibility of being conscious for an artificial system. According to John Searle (1992), for instance, intentionality is an intrinsic property of the human brain, so that being self-aware for an artificial system is simply matter of an arbitrary attributive choice. 'Bodyism' is, in a sense, a rather extreme kind of biological supremacy in current cognitive science. Bodyism is also a way to refer to the 4E cognition, i.e., the embodied, embedded, extended and enacted cognition (Chemero, 2009). All such approaches, according to Manzotti and Chella, are characterized by:

> the confusion between constitution and causation. While they have the resources to defend a causal role of embodiment and environment in shaping consciousness and cognition, they are far from showing convincingly that consciousness and cognition are constituted by the interaction between the body and the environment.
>
> (Manzotti and Chella, 2020, p.191)

Here, the intention behind cognitivism is an ideological one, strongly grounded in computationalism, and for this reason it suffers from all the well-known problems arising from physicalism.

Against all these ideological ways of, it seems preferable to adopt an attitude inspired by pragmatism. "Can computers become conscious or self-conscious?" The best answer is: "It depends." The possibility of machine consciousness depends on the way we deal with some design constraints. It is a matter of accepting the thesis of computational sufficiency, "stating that the right kind of computational structure suffices for the possession of a mind" (Chalmers, 2011, p.326) and to add some computational design constraints. To understand which constraints these are, we need to take into consideration two preliminary ideas and to ask for a key question. First, we have to consider that there are two ways for AI systems to replicate some significant aspects of consciousness: by simulating how this happens in humans, a bio-inspired perspective, and by using deep learning techniques. While there is a naturalizing concern in the first, the second does not worry about the human-like nature of the cognitive architecture which is supposed to generate the feeling of being conscious.

After having produced major advances like vision in the perceptual fields (see §5.3.1), deep learning techniques seem ready to address high level cognitive processes, like consciousness, both in terms of accessibility and phenomenology (Mallakin, 2019). With his Consciousness Prior Hypothesis, Yoshua Bengio (2017) offers a paradigmatic example of this kind of attempt. While most deep learning investigations in cognitive science are focused on unconscious inference over low level inputs, Bengio would like to extend the achievements of deep learning beyond these fields:

> Instead of making predictions in the sensory (e.g., pixel) space, one can thus make predictions in this high level abstract space, which do not have to be limited to just the next time step but can relate events far away from each other in time.

> (Bengio, 2017, p.1)

He uses both an attention mechanism device, able to extract the pertinent features from the background, and a device able to make predictions about the future. As these are two typical cognitive processes involved in self-consciousness, Bengio realized that machine learning and deep learning networks are the best computational architectures available to model self-consciousness.

As remarked above, being self-conscious means two different sorts of things. On the one hand, it means having the ability to recognize oneself from a bodily point of view. On the other, it involves exercising the conceptual and linguistic side of one's inner life, abandoning oneself to the well-known stream of consciousness, which is also the most celebrated part of human self-consciousness. The same applies to artificial self-consciousness. Again, it is a matter of having the two components just mentioned interact in shaping that singular feeling of being present to oneself and representing the world in a reflective and flexible way. This means, inter alia, that in order to speak of artificial self-consciousness in an exhaustive way, there must be a body that can be recognized. In other words, artificial self-consciousness requires robots, which are endowed with a body, and not mere disembodied computers. The above account, therefore, argues for the idea that designing an artificial architecture able to produce the feeling of being a conscious creature implies the capacity to reproduce a twofold cognitive process, made up by a recognition device and an inner speech machinery. The recognition device should include a matching process able to bring together the physical features of the artificial body with those of the observed image. This happens, for instance, in the case of mirror self-recognition (Takeno et al., 2005; Ciliberto et al., 2011; Broun et al., 2014; Zeng et al., 2018).

The machinery for inner speech should be modeled on the above mentioned blueprints. The efforts to model inner speech in a robot are generally based on the idea of turning the ability to hold a speech for public purposes towards oneself. This depends on the conjunction of two theoretical assumptions that are accepted more or less explicitly, namely, on the one hand, some form of social theory of consciousness and, on the other hand, the fact that we have considerably more developed knowledge about how language is produced and understood in the public sphere as opposed to the private, subjective one. Particularly relevant in this field of investigation are the studies carried out by Alain Morin (2006, 2018). A significant attempt to apply what we know about inner speech in terms of artificial modeling is due to Antonio Chella and his colleagues (Chella and Pipitone, 2019; Chella et al., 2020; Chella and Pipitone, 2020), see also (Arrabales, 2012). On the whole, we have to reproduce a twofold cognitive process endowed with a recognition device and an inner speech machinery. Furthermore, we have to keep in mind that self-consciousness is for social purposes. Then, we have to model common sense knowledge and social cognition, and design a cognitive architecture compatible with the mind reading priority account.

7.3   Artificial Morality

If machines are capable of some form of self-consciousness, then the prospect of some sort of moral concern also becomes plausible. In a sense, it is indeed the case that, awareness of how the world works, and also some form of awareness of the role we play in that scene, are conditions of possibility for the development of the capacity to make moral judgments. Morality consists primarily of judgments about observable behavior and its hidden intentions. Because it consists of judgments and intentions, morality seems, at first glance, to be a specifically human creature. Morality exists only where there is human environment, which consists of language, intentional vocabulary, and judgments about the world. Intuitively, then, ethics should only exist in the evaluation of relationships between humans.

Otherwise there should be no ethics, even in the relationship between humans and tools, or with other animals. Or, indeed, with machines.

This kind of reasoning has been challenged by findings in cognitive ethology that show that morality is also a concept that should be studied from an evolutionary perspective. Dutch primatologist Frans de Waal is a champion of ethical consideration of animal behavior (de Waal, 1997, 2006, 2013). He believes that great apes, at least, have feelings of compassion and fairness. Chimpanzees, for example, are accustomed to comforting their conspecifics by hugging and kissing them. This is a significant observation. Indeed, in order to be moved to comfort another individual, one must first understand what they are feeling. Some form of empathy is a prerequisite to being compassionate. If empathy is one of the pillars of morality, then the other is a sense of fairness and reciprocity. This feeling is not lacking in other primates either. De Waal conducted a famous experiment on this with capuchin monkeys. This experiment was later replicated with several other species. Imagine two capuchin monkeys locked in two cages, side by side. One experimenter offers one monkey a piece of cucumber in exchange for a stone, and the monkey finds the exchange beneficial. Then the experimenter offers the second monkey some grapes in the same way. The monkeys like grapes much more than cucumbers. As soon as the first monkey notices the different treatment of the second monkey, he begins to rebel, screaming, banging his fists on the transparent partition, and throwing pieces of cucumber at the experimenter each time he receives them. He seems to feel a sense of injustice and frustration, and feels that there is something unfair about the other monkey receiving better treatment for the same behavior.

Similarly, the trust in reciprocity that arises from a sense of justice also has a counterpart in chimpanzee social behavior. The trust game is an experimental scenario in which an individual is faced with a moral dilemma. The scenario is as follows. One receives a small reward and can keep it all to oneself. In this case, the benefit is immediate and very tangible. Alternatively, one can give one's reward to a second individual, calculating on the other's behavior. In particular, one can hope that the second individual will return some of the reward, since otherwise he would have received nothing. In the second case, one relies on the sense of gratitude and reciprocity that the second individual should cultivate if he shares the same moral framework as we do. But one cannot be sure. After all, nothing prevents the second individual from just taking all the reward. These kinds of tests are administered to people in various ways to measure their propensity to trust others and to understand how the sense of reciprocity works. Using the same type of procedure, we see that chimpanzees also tend to trust and develop a sense of reciprocity. In fact, they go into intersubjective relationships with a fairly high propensity to trust. Chimpanzees are also able to gradually modulate their own trusting behavior based on the behavior of their partner, and to observe whether the partner is behaving more or less trustworthy and correct.

In the case of machines, however, there seems to be an irresolvable contradiction in the ethical consideration of the relation to man. On the one hand, we are interested in building anthropomorphic, intelligent machines because we want to entrust them with a range of tasks that are unpleasant, repetitive, or dangerous for us. On the other hand, we are also interested in giving such machines as human a form as possible, by endowing them with the ability to understand language, to produce appropriate speech in a range of standard situations, to find the right information for a considerable range of states of affairs (from train and flight schedules to weather forecasts, from where nearby to find what we want to buy to where to find a good brunch on Sunday). By endowing machines with such properties, however, they quickly become prime candidates for claiming rights and moral consideration. The same thing happened with other animals. It is the attempt to build human-like machines that creates the problem of their moral and legal consideration when the original intent was to get some sort of artificial slave. Finally, there is no liberation movement for industrial robotics. No one has any real empathy for industrial robots. Imagining what it's like to be a welding machine that tirelessly repeats the same motion all day, welding two pieces of metal together for no apparent purpose, is a case of misplaced empathy. In fact, the welding machine feels nothing, and that is precisely what allows us to use these devices without regard for their (non-existent) inner life.

In the first decades of the artificial intelligence adventure, one of the main objections to the possibility that computers could have real intelligence was to point out to the interlocutor, who was enthusiastic about the technological achievements of computer science, that, after all, the computer wasn't doing anything it hadn't been taught to do before. When confronted with performing a particular skill, typically quickly and automatically, it was pointed out that the knowledge needed to perform that task was provided to the computer before it performed it, albeit in a very slow and mechanical way. In short, computers don't know how to do anything they haven't been taught beforehand. They just know how to do it quickly and automatically, without flexibility or afterthoughts. However, with the success of deep learning, described in §5.2, things have changed dramatically. Deep learning, especially in its unsupervised form, is able to detect significant regularities in a given phenomenon, providing the basis for understanding it. For example, Google's algorithms have begun to blaze a new trail for automatic language understanding. Instead of wracking their brains over what 'understanding' actually means, they focused on complex statistical analyses related to the occurrence of words in the various contexts in which they appear. Such regularities allowed them to make predictions about what role a single term plays in the semantic network in which it occurs, and it often turned out that these regularities and the resulting predictions were all that could be known about the meaning of that word. The possibilities of translation from one language to another thus accelerated dramatically, making the old dream of automatic translation finally possible.

At other times, deep learning is not left alone in trying to detect regularities in the phenomenon it is studying. Goals are set, the learning process is incentivized by rewards and punishments, or examples are provided of what the machine would learn. These machine learning techniques may seem less ambitious than free. But this is just an illusion. On the contrary, if we compare it to the way learning takes place in the animal world and in human education, we can see that deep learning, which is guided in a certain direction by examples, rewards and punishments, is in some ways more ecological than the completely free one. It is a model closer to human and animal learning than a computationally adept machine that is simply left alone with the cyclopean and obscure task of understanding the world before it, with no clear purpose. So, from a period when we worried about the low autonomy of computers, we have arrived at our days when the worry is the opposite, namely the excessive autonomy of computers, which we risk no longer understanding if what they do is done according to a project that is at least initially aligned with humans. This raises the problem of alignment. In the words of Brian Christian:

> As machine-learning systems grow not just increasingly pervasive but increasingly powerful, we will find ourselves more and more often in the position of the "sorcerer's apprentice": we conjure a force, autonomous but totally compliant, give it a set of instructions, then scramble like mad to stop it once we realize our instructions are imprecise or incomplete -lest we get in some clever, horrible way, precisely what we asked for. How to prevent such a catastrophic divergence — how to ensure that these models capture our norms and values, understand what we mean or intend, and, above all, do what we want — has emerged as one of the most central and most urgent scientific questions in the field of computer science. It has a name: the alignment problem.
>
> (Christian, 2020, p.30–31)

Overall, two problems can be distinguished in the field of artificial ethics. The first concerns the question: What are the ethics needed to govern relations with machines that have intelligence? The second type of problem concerns the question: What ethics do machines that are equipped with intelligence and self-awareness have? The first question lacks any ontological commitment to machine morality. It simply concerns the kind of ethics we should adopt to govern our dealings with machines that are assumed to be intelligent, or whose behavior appears to be endowed with some form of intelligence. The story goes as follows. As long as the machines we are dealing with show no signs of intelligence, the question of their ethical evaluation does not arise at all. They are simple tools, like hammers or classic household appliances. A blender, an electric stove, or a refrigerator does not require moral consideration. They are simple tools that we use when we are interested in their functions. But, if at some point, it seems that an appliance needs to make a decision, then that will change. What if blenders were designed to refuse to grind meat? Suppose the designers were vegans and radical animal rights activists, and we found a way to prohibit their blenders from doing to meat what they normally do to vegetables or inert substances. Would we still look at blenders in an ethically neutral way, or wouldn't some of the ethical considerations we apply to designers who advocate for animals also apply to their blenders? So perhaps a vegan activist might feel comfortable with such a blender, while a burger lover would not be pleased to have it in their home. If this kind of reasoning has its own plausibility with a simple blender that may not grind meat, consider how the need for ethics capable of regulating relationships with intelligent machines will become increasingly urgent as AI is distributed in machines.

In addition to two types of problems, there are also two different approaches used to address the above issues: top-down and bottom-up approaches.

> Top-down approaches involve implementing explicit theories about moral behavior in algorithms. Bottom-up approaches involve attempts to train or develop agents whose behavior emulates morally praiseworthy human behavior.
>
> (Allen et al., 2005, p.149)

Top-down approaches start from a pre-existing ethical framework. So let us assume that someone is either a utilitarian or a Kantian committed to a deontological ethic. Based on his or her own biases, the utilitarian or Kantian will try to specify a set of obligations that intelligent machines should obey; in other words, he or she will try to specify constraints that the designers of intelligent machines, who are also moral agents, must obey if they are to design artifacts that are socially acceptable. The goal can be achieved either by building utilitarian or Kantian machines, or by designing machines whose behavior is compatible with utilitarian or Kantian ethics. Bottom-up approaches, on the other hand, emphasize the development of moral sensibility through machines, and take a stance similar to evolutionary theory, imagining the path that machines should take to acquire the moral sensibility that humans have gained through millennia of biological evolution and nurture. Such an approach can be found in developmental robotics, which has moved away from the challenge of developing cognitive skills but is rapidly expanding into the development of moral skills (Cangelosi and Schlesinger, 2015).

The case of autonomous driving cars is an interesting example that lies somewhere between top-down and bottom-up approaches. It also lies somewhere in the middle between approaches that are interested in whether machines really have their own ethics, and those that more modestly ask what ethics we should adopt in order to have fruitful relationships with 'moral' machines. The point is that self-driving cars have choices to make during their operation, and sometimes those choices are ethical in nature. There are different levels of autonomous driving. SAE International (Society of Automotive Engineers) has identified 6 levels, summarized in Figure 7.2. Level 0 (No driving automation): Most vehicles in use

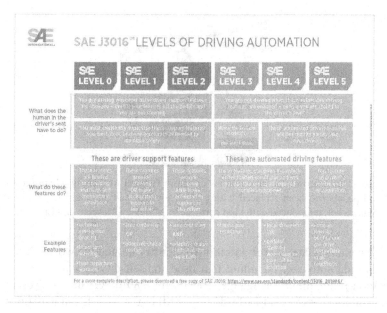

Figure 7.2: SAE International (Society of Automotive Engineers): 6 levels of driving automation.

today belong to level 0, i.e., they are manually controlled; Level 1 (Driver Assistance): The vehicle has a single automated system that assists the driver, e.g., when steering or accelerating (cruise control); Level 2 (Partial Driving Automation): The vehicle is equipped with advanced driver assistance systems (ADAS), e.g., when the vehicle controls both steering and acceleration/deceleration. At this level, the driver can take control of the vehicle at any time. Tesla Autopilot qualifies as Level 2; Level 3 (Conditional Driving Automation): The vehicle has "environment awareness" capabilities and can make decisions such as accelerating past a slow-moving vehicle. The human driver must remain ready to take control if the system cannot perform the task; Level 4 (Highly Automated Driving): At this level, the human driver intervenes if something goes wrong or a system error occurs; Level 5 (Full Driving Automation): The car does not require human attention. Fully autonomous vehicles are currently being tested but are not yet available for the market.

Full Driving Automation may sound like a typical myth of artificial intelligence enthusiasts. But there are reasons to believe that this level of automation in driving could be achieved in the next few years. When Normal Bel Geddes unveiled his design for a radio-controlled vehicle in 1939, no one could have imagined how far vehicle automation would go (Marchand, 1993). General Motors released its first prototype self-driving vehicle in 1958, pictured in Figure 7.3 (Ackerman, 2016). Since then, there have been competitions between quasi-autonomous vehicles built by research teams from different universities or

Figure 7.3: The experimental autonomous car from General Motors, in which the steering wheel and pedals had been replaced with a small joystick and an emergency brake (photo Joseph Scherschel, reproduced with kind permission of Hagley Museum and Library).

research centers related to the automotive industry around the world. Almost all deep learning software, including that used for self-driving cars, is written in the Python programming language, created in 1989 by Guido van Rossum. In 2018, in an interview with Lex Fridman for MIT AI Podcast, van Rossum stated that if machine intelligence emerges, this is most likely to happen in self-driving cars. The problem at this point is no longer to predict exactly when full driving automation will see its full development, but to be socially prepared for such a scenario. The social response to such a scenario is now the most difficult problem to solve for a society where vehicles will do what they need to do without actually being driven by a human.

In the 'social reaction' we must also include the psychological reaction of users who are not yet prepared to be driven around by vehicles that do not need a driver to get where they need to go. It is important to realize that there is something unnatural about the experience of getting around in any car, not only in a self-driven car. No matter how accustomed we are to riding a motorcycle, car, or plane, the old Pleistocene brain we still have in our skull box resists seeing these experiences as truly ecological. Subjecting our brains to additional stress, where we are not only jolted at speeds that were impossible before the industrial revolution, but also carried by vehicles we do not even drive, is a stress still waiting to be overcome. We can no longer make assumptions about the driving styles of people we encounter on the road (Grasso et al., 2019).

Ethics becomes a relevant issue in the field of automotive automation because it is imagined that at some point, as automation progresses, cars will face the same kind of moral dilemma that human drivers sometimes face. The problem is, to be honest, greatly magnified. What comes to mind when talking about ethics and self-driving vehicles is a scenario where the driver has to decide how to act in an emergency situation, where each alternative has its own ethical consequences. For example, imagine the case of a groundhog that suddenly emerges from the woods and enters our lane on a country road. What should we do? Should we run over the groundhog, minimizing the risks to ourselves, our fellow drivers, and others who might be affected by a sudden change in the roadway? Or should we, after all, primarily try not to run over the animal, leaving in the background how we should set up a diversion and its consequences? What makes thinking about such scenarios a somewhat unnatural task is both their relative rarity and the fact that, regardless of the moral stance we would like to take, our driving response is likely to be predominantly due to automaticity and instinctive behavior.

However, ethicists and artificial intelligence scientists have had to work together to develop the right algorithms to allow self-driving vehicles to deal with scenarios similar to that just described. The most popular way to address this issue is to fall back on the celebrated 'trolley dilemma'. Here is its original formulation:

> Suppose that a judge or magistrate is faced with rioters demanding that a culprit be found for a certain crime and threatening otherwise to take their own bloody revenge on a particular section of the community. The real culprit being unknown, the judge sees himself as able to prevent the bloodshed only by framing some innocent person and having him executed. Beside this example is placed another in which a pilot whose aeroplane is about to crash is deciding whether to steer from a more to a less inhabited area. To make the parallel as close as possible, it may rather be supposed that he is the driver of a runaway tram, which he can only steer from one narrow track on to another; five men are working on one track and one man on the other; anyone on the track he enters is bound to be killed.

> (Foot, 1967, p.7)

Several versions of the 'trolley dilemma' have been proposed to explore the question of moral algorithms for regulating self-driving vehicles. The literature is now particularly rich; many scenarios are proposed to measure the moral sentiments prevalent in the test subject group. Some of these scenarios can also be found on the 'Moral Machine' website created by Jean-Frans Bonnefon, one of the leading scholars in this field of research. Consider, for example, the scenario presented by the 'Moral Machine' shown in Figure 7.4. In it, our eventual species preferences are questioned. The same can be done with gender or age preferences.

The idea behind this kind of testing is to make explicit the moral intuitions prevailing among humans about the dilemmas being presented, so that algorithms can later be written to regulate the behavior of cars accordingly. Cars will thus behave by themselves exactly as they would if driven by most humans. This approach precludes the possibility of making cars behave better than humans would. Still, it is not an unrealistic prospect. There have been many studies, the best known of which are those on the so-called 'Lucifer effect', showing that people engage in morally reprehensible behavior under certain circumstances (Zimbardo, 2007). Here we might wish that machines behaved better than humans in the same circumstances. But that is not the case with moral dilemmas

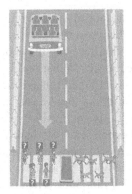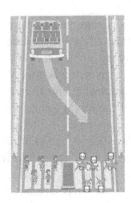

Figure 7.4: Example of dilemmas in the Moral Machine. If you go straight ahead, you may or may not kill five girls. If you skip to the side, five dogs are definitely killed.

in autonomous driving. In the latter case, the goal of scientists is to prevent machines from making decisions that most of us would not make.

However, there is another important variable to consider, and that is the speed with which these types of decisions are made. It is one thing to make a thoughtful decision, it is another to make a sudden decision. Sudden decisions can be influenced by factors we do not want to come into play. In the 2014 Scandinavian film *Turist*, directed by filmmaker Ruben Östlund, an avalanche suddenly heads toward a restaurant terrace where a family is enjoying the spectacle French Alps. Startled by the sudden danger, the children's mother and father behave differently. While the mother instinctively protects her children, the father grabs his gloves and his own cell phone and runs away. The film shows how the shame the father feels for his instinctive behavior eventually corrodes family relationships beyond repair. This is a case of instinctive moral behavior that we do not want to maintain and probably do not want to see turned into an algorithm implemented in a machine. For these kinds of reasons, some researchers have tested what differences, if any, there are in the responses to a moral dilemma in driving by people who have to make a decision very quickly, or by those same people when they are given enough time to think through the circumstances in a thoughtful way. Generally speaking, the results show that while the utilitarian choice had marginal prevalence in the cold task, that is, when you have not to make the decision suddenly, it dominated in the hot task, that is, when the decision must be taken immediately (Bonnefon et al., 2016; Lucifora et al., 2021).

7.4 Future Souls

At the end of this long journey about the various ways in which AI has taken shape in recent years, it becomes natural to ask questions about the more distant future. Given the recent evolution of AI, is it possible to imagine what scenario we will face in the more distant future? A first answer to this question is that we have to be very careful with such predictions. In the first phase of cognitive science, it seemed that the AI of the future would take a form related to computational power and abstract calculations. In a second phase, it seemed that the future should be populated by embodied intelligences, i.e. realized in artificial bodies similar to humans both in terms of external appearance (humanoid robotics) and functional architectures. Recent developments in the field of deep neural networks seem to have changed expectations once again. It now seems that many complex cognitive tasks are within the reach of AI, thanks to functional architectures that work in an associative way, like human brains, but that at some point depart from the human way of thinking and resort to sophisticated mathematical solutions, though no longer ecologically. Inferential statistics and predictive coding seem to predict a future with amazing, but not very human-like results. A humble and cautious attitude is called for in observing these continuous course changes over the last decades.

There is, however, one observation which seems to stand up to the various changes, and which stands out for our thoughts on the future of the artificial mind. However you want to look at the future of AI, it seems that there are two characteristics that you can bet on: Intelligence will be something that is distributed throughout the environment, and something that seems to lose its bearer. The distribution of intelligence can be clearly seen in the proliferation of chips everywhere: from self-driving cars to washing machines, from wearable technology to sex dolls, from watches to drink dispensers. All over the world, we are surrounded by devices that have some cognitive capability embedded in them. Most of these devices do not aspire to realize general intelligence. They are not human-like or other animal-like intelligences. Both we humans and other animals are endowed with a type of intelligence that is predisposed to solving specific tasks that are extremely common in our routines and can therefore be performed automatically and quickly, and without recourse to a cognitively very sophisticated function such as consciousness.

However, the most relevant feature of animal intelligence is that it is able to deal with situations that we have never seen before. It is thanks to this kind of intelligence that in today's world of crowded subways, constant communication with people from all over the world, and cities populated by tens of millions of people, we humans are able to cope with a brain that is still largely the same as the one we had in the Pleistocene. Of course, we are equipped with some automatisms and impulses that we might not want to have today. For example, the overpowering reproductive instinct, driven by all the impulses and compulsions studied by evolutionary psychology, no longer seems so appropriate for individuals who live many decades beyond their reproductive capacity and who, in any case, at least in Western societies, no longer want to have so many children uncontrollably. But even though we carry around unnecessary burdens in today's society that stem from our long evolutionary history, the Pleistocene animal brain is still, by and large, capable of doing its job wonderfully even in third millennium cities.

What is really new, however, is that the mind is no longer just a property of the brains of living things, but something that is found diversely distributed throughout the world. It can be found in a thousand forms. There are humanoid robots or sex dolls that aspire to become fellow humans, capable of empathy and warm relationships similar to those we have with other humans. There are smaller intelligences, so to speak, with a more limited scope that aim to solve specific problems, such as determining when to turn on the lights in a closed environment or when to activate the irrigation system in the garden. Minds are becoming more extended and intelligence more distributed, even if it is not always the same kind of intelligence.

On the other hand, by being distributed everywhere, intelligence runs the risk of losing one of its properties which hitherto seemed essential, namely, the fact of being endowed with a bearer. The point is that intelligence is a property that ultimately concerns mental representations. The latter always have a bearer. The story goes as follows. A certain behavior appears to be endowed with intelligence in our eyes. We ask why this is so, and find that the answer lies in a particular functional architecture that manipulates a set of mental representations in a characteristic way. Both functional architectures and mental representations are things that are owned by someone. In general, the bearer of intelligence can be a non-human animal or a human individual. More recently, we have learned that it can also be an artificial subject, i.e., an abstract computer or a more or less humanoid robot. The subject may be human, animal, or artificial, but in either case it is required by the logical structure of intentional states. There is a theoretical danger, however, if there is a kind of intelligence that has no normal bearer, i.e., something that can be found everywhere but for which no one is responsible. Such intelligence is, so to speak, self-sustaining, as if it were hanging by the laces of its own shoes. Intelligence 'on bootstraps' seems to be a possible future scenario for this crucial property of the mind.

Of course, we may ask what exactly mind and intelligence are without their bearers. Is spread of mind elsewhere the apotheosis of mind or its collapse? Detached from their ordinary users and owners, all this information and computational power run the risk of a 'splodge scenario' rather than one of singularity (Nunn, 2016; Plebe and Perconti, 2020). Or, maybe we run the risk of an unexpected singularity, which does not produce any super-intelligence, but a broken intelligence, that is, a cognitive faculty without anyone who has it, something like an oxymoron.

References

Abadi, M., Agarwal, A., Barham, P., Brevdo, E., Chen, Z. et al. (2015). TensorFlow: Large-scale machine learning on heterogeneous systems. Technical report, Google Brain Team.

Ackerman, E. (2016). Self-driving cars were just around the corner — in 1960. *IEEE Spectrum*, 31 Aug.

Adiwardana, D., Luong, M. -T., So, D. R., Hall, J., Fiedel, N., Thoppilan, R. et al. (2020). Towards a human-like open-domain chatbot. *arXiv*, abs/2001.09977.

Admoni, H. and Scassellati, B. (2017). Social eye gaze in human-robot interaction: A review. *Journal of Human-Robot Interaction*, 6: 25–63.

Agafonova, Y., Tikhonov, A. and Yamshchikov, I. P. (2020). Paranoid Transformer: Reading narrative of madness as computational approach to creativity. *Future Internet*, 12: 1–12.

Agre, P. E. and Chapman, D. (1987). Pengi: An implementation of a theory of activity. In *Proceedings of the Sixth Conference on Artificial Intelligence*, pages 268–272, Seattle (WA).

Aldrich, R. J. (2010). *GCHQ: The Uncensored Story of Britain's Most Secret Intelligence Agency*. HarperCollins, London.

Allen, C., Smit, I. and Wallach, W. (2005). Artificial morality: Top-down, bottom-up, and hybrid approaches. *Ethics and Information Technology*, 7: 149–155.

Allen, J. (1995). *Natural Language Understanding*. Benjamin Cummings, Philadelphia.

Anderson, J. A. and Rosenfeld, E., editors (2000). *Talking Nets: An Oral History of Neural Networks*. MIT Press, Cambridge (MA).

Andreopoulos, A. and Tsotsos, J. K. (2013). 50 years of object recognition: Directions forward. *Computer Vision and Image Understanding*, 117: 827–891.

Arbilly, M. and Lotem, A. (2017). Constructive anthropomorphism: A functional evolutionary approach to the study of human-like cognitive mechanisms in animals. *Proceedings of the Royal Society of London B*, 284: 20171616.

Aristotle (335-323BCE). *Organon*. English translation by Octavius Freire Owen, 1853, Henry Bohn, London.

Arrabales, R. (2012). Inner speech generation in a video game non-player character: From explanation to self? *International Journal of Machine Consciousness*, 4: 367–381.

Arrow, K. J. (2005). John Holland and the evolution of economics. In Booker, L., Forrest, S., Mitchell, M. and Riolo, R., editors, *Perspectives on Adaptation in Natural and Artificial Systems*, pages 281–290. Oxford University Press, Oxford (UK).

Ashby, W. R. (1947). Principles of the self-organizing dynamic system. *The Journal of General Psychology*, 37: 125–128.

Ashby, W. R. (1949). Experimental Homeostat. *Electroencephalography and Clinical Neurophysiology*, 1: 116–117.

Ashby, W. R. (1962). Principles of the self-organizing system. In von Foerster, H. and Zopf, G. W., editors, *Principles of Self-Organization: Transactions of the University of Illinois Symposium*, pages 255–278. Pergamon Press, New York.

Asimov, I. (1956). *First Law*. Fantastic Universe, New York. Collected in *The Rest of the Robots*, Doubleday, New York, 1964.

Avenanti, A., Bueti, D., Galati, G., and Aglioti, S. M. (2005). Transcranial magnetic stimulation highlights the sensorimotor side of empathy for pain. *Nature Neuroscience*, 8: 955–960.

Babbage, C. (1889). *Babbage's Calculating Engines – Being a Collection of Papers Relating to Them; Their History, and Construction*. E. and F.N. Spon, London. Edited by Henry P. Babbage.

Backus, J. W. (1959). The syntax and semantics of the proposed international algebraic language of the Zurich ACM-GAMM conference. In *Proceedings of International Conference on Information Processing*, pages 125–132. UN-ESCO.

Baddeley, A. (1992). Working memory. *Science*, 255: 556–559.

Baddeley, A. D. (1967). How does acoustic similarity influence short-term memory? *Quarterly Journal of Experimental Psychology*, 20: 249–263.

Baddeley, A. D. and Hitch, G. (1974). Working memory. In Bower, G., editor, *The Psychology of Learning and Motivation: Advances in Research and Theory*. Academic Press, New York.

Barkow, J. H., Cosmides, L. and Tooby, J., editors (1995). *The Adapted Mind: Evolutionary Psychology and the Generation of Culture*. Oxford University Press, Oxford (UK).

Barrett, P. H. (1987). *Charles Darwin's Notebooks, 1836-1844: Geology, Transmutations of Species, Metaphysical Enquiries*. Cornell University Press, Ithaca (NJ).

Bartunov, S., Santoro, A., Richards, B. A., Marris, L., Hinton, G. E., and Lillicrap, T. (2018). Assessing the scalability of biologically-motivated deep learning algorithms and architectures. In *Advances in Neural Information Processing Systems*.

Bauer, F. L. (2010). *Origins and Foundations of Computing*. Springer-Verlag, Berlin.

Bayes, T. (1763). An essay towards solving a problem in the doctrine of chances. *Philosophical Transactions of the Royal Society of London*, 53: 370–4185.

Beaver, P. (1974). *Victorian Parlor Games*. Thomas Nelson, Edinburgh.

Bednar, J. A. (2002). *Learning to See: Genetic and Environmental Influences on Visual Development*. PhD thesis, University of Texas at Austin. Tech Report AI-TR-02-294.

Bednar, J. A., Choe, Y., Paula, J. D., Miikkulainen, R., Provost, J. and Tversky, T. (2004). Modeling cortical map with Topographica. *Neurocomputing*, 58-60: 1129–1135.

Bellemare, M. G., Naddaf, Y., Veness, J. and Bowling, M. (2013). The Arcade learning environment: An evaluation platform for general agents. *Journal of Artificial Intelligence Research*, 47: 253–279.

Belousov, B. (1959). Periodically acting reaction and its mechanism. *Collection of Abstracts on Radiation Medicine*, 147: 145. Originale in lingua russa.

Bénard, H. (1900). Les tourbillons cellulaires dans une nappe liquide. *Revue Générale des Sciences*, 11: 1261–1271, 1309–1328.

Bender, E. M. and Koller, A. (2020). Climbing towards NLU: On meaning, form, and understanding in the age of data. In *58th Annual Meeting of the Association for Computational Linguistics*, pages 5185–5198, Somerset (NJ). Association for Computational Linguistics.

Bengio, Y. (2017). The consciousness prior. *arXiv*, abs/1709.08568.

Bengio, Y., Lamblin, P., Popovici, D. and Larochelle, H. (2007). Greedy layerwise training of deep networks. In *Advances in Neural Information Processing Systems*, pages 153–160.

Benveniste, A., Metivier, M. and Priouret, P. (1990). *Adaptive Algorithms and Stochastic Approximations*. Springer-Verlag, Berlin.

Bernoulli, D. (1738). Exposition of a new theory on the measurement of risk. *Econometrica*, 22: 23–36.

Bernoulli, J. (1713). *Ars Conjectandi*. Impensis Thurnisiorum, fratrum, Basel.

Betti, E. (1872). Il nuovo cimento. *Series*, 2: 7.

Beurle, R. L. (1956). Properties of a mass of cells capable of regenerating pulses. *Philosophical Transactions of the Royal Society B*, s40: 55–94.

Beurle, R. L. (1962). Storage and manipulation of information in random networks. In Muses, C. A., editor, *Aspects of the Theory of Artificial Intelligence*, pages 19–42. Springer-Verlag, Berlin.

Bhattacharya, S., Maddikunta, P. K. R., Pham, Q.-V., Gadekallu, T. R., S, S. R. K., Chowdhary, C. L., Alazab, M. and Piran, M. J. (2021). Deep learning and medical image processing for coronavirus (COVID-19) pandemic: A survey. *Sustainable Cities and Society*, 65: 102589.

Bhaumik, A. (2018). *From AI to Robotics - Mobile, Social, and Sentient Robots*. CRC Press, Boca Raton (FL).

Bianchini, M. and Scarselli, F. (2014a). On the complexity of neural network classifiers: A comparison between shallow and deep architectures. *IEEE Transactions on Neural Networks and Learning Systems*, 25: 1553–1565.

Bianchini, M. and Scarselli, F. (2014b). On the complexity of shallow and deep neural network classifiers. In *Proceedings of European Symposium on Artificial Neural Networks*, pages 371–376.

Bickle, J. (2003). *Philosophy and Neuroscience: A Ruthlessly Reductive Approach*. Kluwer, Dordrecht (NL).

Blanchard, P. and Brüning, E. (1992). *Variational Methods in Mathematical Physics: A Unified Approach*. Springer-Verlag, Berlin.

Block, N., editor (1981a). *Imagery*. Cambridge University Press, Cambridge (UK).

Block, N. (1981b). Psychologism and behaviorism. *Philosophical Review*, 90: 5–43.

Bo, L., Lai, K., Ren, X. and Fox, D. (2011). Object recognition with hierarchical kernel descriptors. In *Proceedings of IEEE International Conference on Computer Vision and Pattern Recognition*, pages 1729–1736.

Boden, M. (2000). Autopoiesis and life. *Cognitive Science Quarterly*, 1: 117–145.

Boden, M. (2008). *Mind as Machine: A History of Cognitive Science*. Oxford University Press, Oxford (UK).

Bojanowski, P., Grave, E., Joulin, A. and Mikolov, T. (2017). Enriching word vectors with subword information. *Transactions of the Association for Computational Linguistics*, 5: 135–146.

Bojar, O., Buck, C., Federmann, C., Haddow, B., Koehn, P., Leveling, J., Monz, C. et al. (2014). Findings of the 2014 workshop on statistical machine translation. In *Proceedings of the Workshop on Statistical Machine Translation*, pages 12–58.

Bonnefon, J.-F., Shariff, A. and Rahwan, I. (2016). The social dilemma of autonomous vehicles. *Science*, 352: 1574–1576.

Boole, G. (1854). *An Investigation of the Laws of Thought, on which are founded the Mathematical Theories of Logic and Probabilities*. Walton and Maberley, London. Trad. it. di Mario Trinchero *Indagine sulle leggi del pensiero, cui sono fondate le teorie matematiche della logica e della probabilita*, 1976.

Bottou, L. and LeCun, Y. (2004). Large scale online learning. In *Advances in Neural Information Processing Systems*, pages 217–224.

Bovens, L. and Hartmann, S. (2003). *Bayesian Epistemology*. Oxford University Press, Oxford (UK).

Bower, J. M. and Beeman, D. (1998). *The Book of Genesis: Exploring Realistic Neural Models with the General Neural Simulation System*. Springer-Verlag, New York, second edition.

Bracewell, R. (2003). *Fourier Analysis and Imaging*. Springer-Verlag, Berlin.

Brandom, R. B. (2000). *Articulating Reasons: An Introduction to Inferentialism*. Harvard University Press, Cambridge (MA).

Brandom, R. B. (2019). *A Spirit of Trust: A Reading of Hegel's Phenomenology*. Harvard University Press, Cambridge (MA).

Brandon, R. (1990). *Adaptation and Environment*. Princeton University Press, Princeton (NJ).

Brandon, R. (1996). *Concepts and Methods in Evolutionary Biology*. Cambridge University Press, Cambridge (UK).

Braun, J., Koch, C. and Davis, J. L. (2001). *Visual Attention and Cortical Circuits*. MIT Press, Cambridge (MA).

Brazier, M. (1961). *A History of the Electrical Activity of the Brain: The First Half-century*. Macmillan, New York.

Brdar, M., Gries, S. and Fuchs, M., editors (2011). *Cognitive Linguistics – Convergence and Expansion*. John Benjamins, Amsterdam.

Bridgman, P. W. (1927). *The Logic of Modern Physics*. Macmillan, New York.

Bridgman, P. W. (1950). The operational aspect of meaning. *Synthese*, 8: 251–259.

Brindley, G. S. and Lewin, W. (1968). The sensations produced by electrical stimulation of the visual cortex. *Journal of Neurophysiology*, 196: 479–493.

Bronkhorst, A. W. (2015). The cocktail-party problem revisited: Early processing and selection of multi-talker speech. *Attention, Perception, & Psychophysics*, 77: 1465–1487.

Brooks, R. A. (1986). A robust layered control system for a mobile robot. *IEEE Journal on Robotics and Automation*, 2: 14–23.

Brooks, R. A. (1990a). The behavior language; User's guide. Technical Report AI Memo 1227, MIT Press.

Brooks, R. A. (1990b). Elephants don't play chess. *Robotics and Autonomous Systems*, 6: 3–15.

Brooks, R. A. (1991). Intelligence without representation. *Artificial Intelligence*, 47: 139–159.

Brooks, R. A. (1997). From earwigs to humans. *Robotics and Autonomous Systems*, 20: 291–304.

Brooks, R. A. (1999). *Cambrian Intelligence: The Early History of the New AI.* MIT Press, Cambridge (MA).

Broun, A., Beck, C., Pipe, T., Mirmehdi, M. and Melhuish, C. (2014). Bootstrapping a robot's kinematic model. *Robotics and Autonomous Systems*, 62: 330–339.

Brown, T. B., Mann, B., Ryder, N., Subbiah, M., Kaplan, J. et al. (2020). Language models are few-shot learners. *arXiv*, abs/2005.14165.

Brown, T. L. (2003). *Making Truth: Metaphor in Science.* University of Illinois Press, Urbana (IL).

Bruni, D., Perconti, P. and Plebe, A. (2018). Anti-anthropomorphism and its limits. *Frontiers in Psychology*, 9: 2205.

Cadieu, C. F., Hong, H., Yamins, D. L. K., Pinto, N., Ardila, D., Solomon, E. A., Majaj, N. J. and DiCarlo, J. J. (2014). Deep neural networks rival the representation of primate IT cortex for core visual object recognition. *PLoS Computational Biology*, 10: e1003963.

Cangelosi, A. and Parisi, D. (1998). The evolution of a 'language' in an evolving population of neural nets. *Connection Science*, 10: 83–97.

Cangelosi, A. and Schlesinger, M. (2015). *Developmental Robotics; From Babies to Robots.* Cambridge University Press, Cambridge (UK).

Cantor, G. (1891). Über eine elementare frage der mannigfaltigketislehre. *Jahresbericht der Deutschen Mathematiker-Vereinigung*, 1: 75–78.

Ĉapek, K. (1920). *R.U.R. – Rossumovi Univerzální Roboti.* Kr. Vinohrady, Praha. English translation by Selver P. and Playfair N. in *Rossum's Universal Robots*, Samuel French, New York, 1923.

Carey, S. and Spelke, E. (1994). Domain-specific knowledge and conceptual change. In Hirschfeld, L. A. and Gelman, S. A., editors, *Mapping the Mind: Domain Specificity in Cognition and Culture*, pages 169–200. Cambridge University Press, Cambridge (UK).

Carey, S. and Spelke, E. (1996). Science and core knowledge. *Journal of Philosophy of Science*, 63: 515–533.

Carrasco, M. (2011). Visual attention: The past 25 years. *Vision Research*, 51: 1484–1525.

Carruthers, P., Fletcher, L. and Ritchie, B. (2012). The evolution of self-knowledge. *Philosophical Topics*, 40: 13–37.

Cauchy, A. -L. (1847). Méthode générale pour la résolution des systèmes d'équations simultanées. *Comptes rendus des séances de l'Académie des sciences de Paris*, 25: 536–538.

Ceaser, J. (1997). *Reconstructing America: The Symbol of America in Modern Thought*. Yale University Press, New Haven (CO).

Ceruzzi, P. E. (2003). *A History of Modern Computing*. MIT Press, Cambridge (MA). 2nd edition.

Chalmers, D. (2010). The singularity: A philosophical analysis. *Journal of Consciousness Studies*, 17: 7–65.

Chalmers, D. (2011). A computational foundation for the study of cognition. *Journal of Consciousness Studies*, 12: 323–357.

Chapman, D. and Agre, P. E. (1986). Abstract reasoning as emergent from concrete activity. In *Proceedings of the Reasoning about Actions and Plans Workshop*, pages 411–424, Los Altos (CA). Morgan Kaufman.

Chatfield, K., Simonyan, K., Vedaldi, A. and Zisserman, A. (2014). Return of the devil in the details: Delving deep into convolutional nets. *arXiv*, abs/1405.3531.

Chella, A. and Pipitone, A. (2019). The inner speech of the IDyOT: Comment on 'Creativity, information, and consciousness: The information dynamics of thinking' by Geraint A. Wiggins. *Physics of Life Reviews*, S1571-S0645: 30024–30027.

Chella, A. and Pipitone, A. (2020). A cognitive architecture for inner speech. *Cognitive Systems Research*, 59: 287–292.

Chella, A., Pipitone, A., Morin, A., and Racy, F. (2020). Developing self-awareness in robots via inner speech. *Frontiers in Robotics and AI*, 7: article 16.

Chemero, A. (2009). *Radical Embodied Cognitive Science*. MIT Press, Cambridge (MA).

Chen, T., Li, M., Li, Y., Lin, M., Wang, N., Wang, M., Xiao, T., Xu, B., Zhang, C. and Zhang, Z. (2015). MXNet: A flexible and efficient machine learning library for heterogeneous distributed systems. *arXiv*, abs/1512.01274.

Chevalier, P., Kompatsiari, K., Ciardo, F. and Wykowska, A. (2020). Examining joint attention with the use of humanoid robots – A new approach to study fundamental mechanisms of social cognition. *Psychonomic Bulletin & Review*, 27: 217–236.

Cho, K., van Merriënboer, B., Gulcehre, C., Bahdanau, D., Bougares, F., Schwenk, H. and Bengio, Y. (2014). Learning phrase representations using RNN encoder–decoder for statistical machine translation. In *Conference on Empirical Methods in Natural Language Processing*, pages 1724–1734. Association for Computational Linguistics.

Chomsky, N. (1956). Three models for the description of languages. *IRE Transaction on Information Theory*, 2: 113–124.

Chomsky, N. (1957). *Syntactic Structures*. Mouton & Co., The Hague (NL).

Chomsky, N. (1958). On certain formal properties of grammars. *Information and Control*, 1: 91–112.

Chomsky, N. (1959). Review of B.F. Skinner's 'Verbal Behavior'. *Language*, 35: 26–58.

Chomsky, N. (1966). *Cartesian Linguistics: A Chapter in the History of Rationalist Thought*. Harper and Row Pub. Inc, New York.

Chomsky, N. (1975). *The Logical Structure of Linguistic Theory*. Springer-Verlag, New York.

Chomsky, N. (1981). *Lectures in Government and Binding*. Foris, Dordrecht.

Chomsky, N. (1993). A minimalist progam for linguistic theory. In Hale, K. and Keyser, S. J., editors, *The View from Building 20: Essays in Linguistics in Honor of Sylvain Bromberger*. MIT Press, Cambridge (MA).

Chomsky, N. (1995). *The Minimalist Program*. MIT Press, Cambridge (MA).

Chomsky, N. (2000). *New Horizons in the Study of Language and Mind*. Cambridge University Press, Cambridge (UK).

Chomsky, N. and Halle, M. (1968). *The Sound Pattern of English*. Harper and Row Pub. Inc, New York.

Chomsky, N. and Lasnik, H. (1993). The theory of principles and parameters. In Jacobs, J., von Stechow, A., Sternefeld, W. and Vennemann, T., editors, *Syntax – An International Handbook of Contemporary Research*. W. de Gruyter, Berlin.

Chomsky, N. and Miller, G. A. (1963). Formal properties of grammars. In Luce, R. D., Bush, R. R. and Galanter, E., editors, *Handbook of Mathematical Psychology*. John Wiley, New York.

Chomsky, N. and Schützenberger, M.-P. (1959). The algebraic theory of context-free languages. *Studies in Logic and the Foundations of Mathematics*, 26: 118–161.

Christian, B. (2020). *The Alignment Problem: Machine Learning and Human Values*. W. W. Norton & Company, New York.

Chui, M., Manyika, J., Miremadi, M., Henke, N., Chung, R., Nel, P. and Malhotra, S. (2018). Notes from the AI frontier: Insights from hundreds of use cases. Technical Report April, McKinsey Global Institute.

Church, A. (1936). An unsolvable problem of elementary number theory. *American Journal of Mathematics*, 58: 345–363.

Church, A. (1937). A. M. Turing. On computable numbers, with an application to the Entscheidungsproblem. Proceedings of the London Mathematical Society, 2 s. vol. 42 (1936–1937), pp. 230–265. *The Journal of Symbolic Logic*, 2: 42–43.

Churchland, P. S. and Sejnowski, T. (1994). *The Computational Brain*. MIT Press, Cambridge (MA).

Churchland, P. S. and Sejnowski, T. J. (1990). Neural representation and neural computation. *Philosophical Perspectives*, 4: 343–382.

Cicchetti, D. V. (1991). The reliability of peer review for manuscript and grant submissions: A cross-disciplinary investigation. *Behavioral and Brain Science*, 14: 119–186.

Cichy, R. M., Khosla, A., Pantazis, D., Torralba, A. and Oliva, A. (2016). Comparison of deep neural networks to spatio-temporal cortical dynamics of human visual object recognition reveals hierarchical correspondence. *Scientific Reports*, 6: 27755.

Ciliberto, C., Smeraldi, F., Natale, L. and Metta, G. (2011). Online multiple instance learning applied to hand detection in a humanoid robot. In *IEEE/RSJ International Conference on Intelligent Robots and Systems*, pages 1526–1532.

Cinbis, R. G., Verbeek, J. and Schmid, C. (2012). Segmentation driven object detection with Fisher vectors. In *International Conference on Computer Vision*, pages 2968–2975.

Cireşan, D., Meier, U. and Schmidhuber, J. (2012). Multi-column deep neural networks for image classification. In *Proceedings of IEEE International Conference on Computer Vision and Pattern Recognition*.

Clarke, A., Devereux, B. J., Randall, B. and Tyler, L. K. (2015). Predicting the time course of individual objects with MEG. *Cerebral Cortex*, 25: 3602–3612.

Coates, A., Huval, B., Wang, T., Wu, D. J., Ng, A. Y. and Catanzaro, B. (2013). Deep learning with COTS HPC systems. In *International Conference on Machine Learning*, pages 1337–1345.

Coghlan, S., Vetere, F., Waycott, J. and Neves, B. B. (2019). Could social robots make us kinder or crueller to humans and animals? *International Journal of Social Robotics*, 11: 741–751.

Collier, B. and MacLachlan, J. (1998). *Charles Babbage and the Engines of Perfection*. Oxford University Press, Oxford (UK).

Collins, H. and Evans, R. (2014). Quantifying the tacit: The imitation game and social fluency. *Sociology*, 48: 3–19.

Collins, O. F. and Moore, D. G. (1964). *The Enterprising Man*. Michigan State University Press, East Lansing (MI).

Conway, B. R. (2018). The organization and operation of inferior temporal cortex. *Annual Review of Vision Science*, 4: 19.1–19.22.

Cooper, S., Daniel, P. M. and Whitteridge, D. (1953). Nerve impulses in the brainstem and cortex of the goat. spontaneous discharges and responses to visual and other afferent stimuli. *Journal of Physiology*, 120: 514–527.

Copeland, J. (2000). The Turing test. *Minds and Machines*, 10: 519–539.

Copeland, J. and Proudfoot, D. (1996). On Alan Turing's anticipation of connectionism. *Synthese*, 108: 361–377.

Copeland, J. and Proudfoot, D. (2004). The computer, artificial intelligence, and the Turing test. In Teuscher (2004), pages 317–351.

Cordeschi, R. (2002). *The Discovery of the Artificial – Behavior, Mind and Machines Before and Beyond Cybernetics*. Springer-Verlag, Berlin.

Cornish-Bowden, A. and Cárdenas, M. L. (2020). Contrasting theories of life: Historical context, current theories. in search of an ideal theory. *BioSystems*, 188: 104063.

Cosmides, L. and Tooby, J. (1997). The modular nature of human intelligence. In Scheibel, A. and Schopf, J. W., editors, *The origin and evolution of intelligence*. Oxford University Press, Oxford (UK).

Cottrell, G. W. and Plunkett, K. (1994). Acquiring the mapping from meaning to sounds. *Connection Science*, 6: 379–412.

Coursey, K. (2020). Speaking with Harmony: Finding the Right Thing to Do or Say While in Bed (or Anywhere Else). In Bendel, O., editor, *Maschinenliebe*, pages 35–51. Springer-Verlag, Berlin.

Cowan, N. and Wood, N. L. (1997). Cog as a thought experiment. *Consciousness and Cognition*, 6: 182–203.

Crane, H. (1960). The Neuristor. *IRE Transactions on Electronic Computers*, 9: 370–371.

Croft, W. and Cruse, A. (2004). *Cognitive Linguistics*. Cambridge University Press, Cambridge (UK).

Cummins, R. (2000). "How does it work?" versus "what are the laws?": Two conceptions of psychological explanation. In Keil, F. C. and Wilson, R. A., editors, *Explanation and Cognition*, pages 117–144. MIT Press, Cambridge (MA).

Curry, H. B. (1944). The method of steepest descent for non-linear minimization problems. *Quarterly of Applied Mathematics*, 2: 258–261.

Cybenko, G. (1989). Approximation by superpositions of a sigmoidal function, mathematics of control. *Signals and Systems*, 2: 303–314.

Dale, R., Moisl, H. and Somers, H., editors (2000). *Handbook of Natural Language Processing*. Marcel Dekker, New York.

Daniel, H.-D. (2005). Publications as a measure of scientific advancement and of scientists' productivity. *Learned Publishing*, 18: 143–148.

Darwin, C. (1859). *On the Origin of Species by Means of Natural Selection*. John Murray, London, first edition.

Darwin, C. (1871). *The Descent of Man, and Selection in Relation to Sex*. John Murray, London.

Darwin, C. (1872). *The Expression of the Emotions in Man and Animals*. John Murray, London.

Darwin, C. (about 1880). Notebooks. Published in Barrett (1987).

Dasgupta, S., editor (2014). *It Began with Babbage: The Genesis of Computer Science*. Oxford University Press, Oxford (UK).

Daugherty, K. and Seidenberg, M. (1992). Rules or connections? The past tense revisited. In Kruschke, J. K., editor, *Proceedings of the XIV Annual Conference of the Cognitive Science Society*, pages 259–264, Mahwah (NJ). Lawrence Erlbaum Associates.

Daugman, J. G. (1990). Brain metaphor and brain theory. In Schwartz, E. L., editor, *Computational Neuroscience*, pages 23–36. MIT Press, Cambridge (MA).

Davidson, D. (2001). *Subjective, Intersubjective, Objective*. Oxford University Press, Oxford (UK).

Davis, M. (2000). *Engines of Logic – Mathematicians and the Origin of the Computer*. Cambridge University Press, Cambridge (UK).

Dawkins, R. (1976). *The Selfish Gene*. Oxford University Press, Oxford (UK).

Dayan, P. and Abbott, L. F. (2001). *Theoretical Neuroscience*. MIT Press, Cambridge (MA).

Daylight, E. G. (2015). Towards a historical notion of 'Turing–the father of computer science'. *History and Philosophy of Logic*, 36: 205–228.

de Finetti, B. (1974). *Theory of Probability*. John Wiley, New York.

de La Mettrie, J. O. (1748). *L'Homme Machine*. Elie Luzac, Leyden.

de Moivre, A. (1718). *The Doctrine of Chances: or, A Method of Calculating the Probability of Events in Play*. W. Pearson, London.

de Villers, J. and Barnard, E. (1992). Backpropagation neural nets with one and two hidden layers. *IEEE Transactions on Neural Networks*, 4: 136–141.

de Waal, F. B. M. (1997). *Good Natured. The Origins of Right and Wrong in Humans and Other Animals*. Cambridge University Press, Cambridge (UK).

de Waal, F. B. M. (2006). *Primates and Philosophers. How Morality Evolved*. Princeton University Press, Princeton (NJ).

de Waal, F. B. M. (2013). *The Bonobo and the Atheist. In Search of Humanism Among the Primates*. W. W. Norton & Company, New York.

Deb, K. and Agrawal, R. (1994). Simulated binary crossover for continuous search space. *Complex Systems*, 9: 1–15.

Dennett, D. C. (1978). *Brainstorms*. Bradford Books, Montgomery (VE).

Dennett, D. C. (1980). The milk of human intentionality. *Behavioral and Brain Science*, 3: 429–430.

Dennett, D. C. (1985). Can machines think? In Shafto, M. G., editor, *How We Know: Nobel Conference XX*. Harper and Row Pub. Inc, New York.

Dennett, D. C. (1987). *The Intentional Stance*. MIT Press, Cambridge (MA).

Dennett, D. C. (1991). Review of C. McGinn, The Problem of Consciousness. *The Times Literary Supplement*, May 10: 10.

Dennett, D. C. (1995). *Darwin's Dangerous Idea. Evolution and the Meaning of Life*. Simon and Schuster, New York.

Dennett, D. C. (1997). Cog as a thought experiment. *Robotics and Autonomous Systems*, 20: 251–256.

Dennett, D. C. (2012). *Intuition Pumps and Other Tools for Thinking*. W. W. Norton & Company, New York.

Dennett, D. C. (2017). *From Bacteria to Bach and Back: The Evolution of Mind*. W. W. Norton & Company, New York.

Deo, N. (1974). *Graph Theory with Applications to Engineering and Computer Science*. Prentice Hall, Englewood Cliffs (NJ).

Deutsch, K. W. (1966). *The Nerves of Government: Models of Political Communication and Control*. Free Press, New York.

Devlin, J., Chang, M.-W., Lee, K. and Toutanova, K. (2019). BERT: Pre-training of deep bidirectional transformers for language understanding. In *Proceedings North American Chapter of the Association for Computational Linguistics: Human Language Technologies*, pages 4171–4186. Association for Computational Linguistics.

Dinan, E., Logacheva, V., Malykh, V., Miller, A., Shuster, K., Urbanek, J., Kiela, D., Szlam, A., Serban, I., Lowe, R., Prabhumoye, S., Black, A. W., Rudnicky, A., Williams, J., Pineau, J., Burtsev, M. and Weston, J. (2020). The second conversational intelligence challenge (convAI2). In Escalera and Herbrich (2020), pages 187–202.

Dollens, D. (2014). Alan Turing's drawings, autopoiesis and can buildings think? *Leonardo*, 47: 249–253.

Dorffner, G., Hentze, M. and Thurner, G. (1996). A connectionist model of categorization and grounded word learning. In Koster, C. and Wijnen, F., editors, *Proceedings of the Groningen Assembly on Language Acquisition*.

Dorigo, M., Maniezzo, V. and Colorni, A. (1996). The ant system: Optimization by a colony of cooperating agents. *IEEE Trans. Syst, Man, Cybern. B*, 26: 29–41.

Douglas, R. J. and Martin, K. A. (2004). Neuronal circuits of the neocortex. *Annual Review of Neuroscience*, 27: 419–451.

Douglas, R. J., Martin, K. A. and Whitteridge, D. (1989). A canonical microcircuit for neocortex. *Neural Computation*, 1: 480–488.

Dow, J. M. (2012). Mindreading, mindsharing, and the origins of self-consciousness. *Philosophical Topics*, 40: 39–70.

Downes, S. M. (2005). Integrating the multiple biological causes of human behavior. *Biology and Philosophy*, 20: 177–190.

Dreyfus, H. (1965). Alchemy and artificial intelligence. Technical Report P-3244, RAND Corporation, Santa Monica (CA).

Dreyfus, H. (1972). *What Computers Can't Do: A Critique of Artificial Reason*. Harper and Row Pub. Inc, New York.

Dreyfus, H. (1990). *Being-in-the-World: A Commentary on Heidegger's Being and Time, Division I*. MIT Press, Cambridge (MA).

Dreyfus, H. (1992). *What Computers Still Can't Do: A Critique of Artificial Reason*. MIT Press, Cambridge (MA).

Dreyfus, H. (2007). Why Heideggerian AI failed and how fixing it would require making it more Heideggerian. *Philosophical Psychology*, 20: 247–268.

Duchi, J., Hazan, E. and Singer, Y. (2011). Adaptive subgradient methods for online learning and stochastic optimization. *Journal of Machine Learning Research*, 12: 2121–2159.

Dummett, M. A. (1973). *Frege: Philosophy of Language*. Duckworth, London.

Duncan, J. (2006). EPS mid-career award 2004 – Brain mechanisms of attention. *Quarterly Journal of Experimental Psychology*, 59: 2–27.

Dupuy, J.-P. (1994). *Aux origines des sciences cognitives*. La Découverte, Paris. English translation by M. B. DeBevoise in *On the Origins of Cognitive Science: The Mechanization of the Mind*. Princeton University Press, 2000.

Durrani, N., Haddow, B., Koehn, P. and Heafield, K. (2014). Edinburgh's phrase-based machine translation systems for WMT-14. In *Proceedings of the Workshop on Statistical Machine Translation*, pages 97–104.

Earley, J. (1970). An efficient context-free parsing algorithm. *Communications of the Association for Computing Machinery*, 13: 94–102.

Earman, J. (1992). *Bayes or Bust? A Critical Examination of Bayesian Confirmation Theory*. MIT Press, Cambridge (MA).

Eberhart, R. C. and Kennedy, J. (1995). A new optimizer using particle swarm theory. In *Proceedings of Sixth International Symposium on Micro Machine and Human Science*, volume 1, pages 39–43.

Eickenberg, M., Gramfort, A., Varoquaux, G. and Thirion, B. (2017). Seeing it all: Convolutional network layers map the function of the human visual system. *NeuroImage*, 152: 184–194.

Eldan, R. and Shamir, O. (2016). The power of depth for feedforward neural networks. *Journal of Machine Learning Research*, 49: 1–34.

Elkins, K. and Chun, J. (2020). Can GPT-3 pass a writer's Turing test? *Journal of Cultural Analytics*, 10.22148/001c.17212.

Elman, J. L. (1990). Finding structure in time. *Cognitive Science*, 14: 179–221.

Elman, J. L. (1991). Distributed representations, simple recurrent networks, and grammatical structure. *Machine Learning*, 7: 195–225.

Elman, J. L. (1992). Grammatical structure and distributed representations. In Davis, S., editor, *Connectionism: Theory and Practice*. Oxford University Press, Oxford (UK).

Elman, J. L. (1993). Learning and development in neural networks: The importance of starting small. *Cognition*, 48: 71–99.

Elman, J. L. (1995). Language as a dynamical system. In Port, R. F. and Gelder, T. v., editors, *Mind as Motion: Explorations in the Dynamics of Cognition*, pages 195–225. MIT Press, Cambridge (MA).

Elman, J. L., Bates, E., Johnson, M. H., Karmiloff-Smith, A., Parisi, D. and Plunkett, K. (1996). *Rethinking Innateness: A Connectionist Perspective on Development*. MIT Press, Cambridge (MA).

Enderton, H. B. (1977). *Elements of Set Theory*. Academic Press, New York.

Epstein, R., Roberts, G. and Beber, G., editors (2008). *Parsing the Turing Test: Philosophical and Methodological Issues in the Quest for the Thinking Computer*. Springer-Verlag, Berlin.

Escalera, S. and Herbrich, R., editors (2020). *The NeurIPS '18 Competition – From Machine Learning to Intelligent Conversations*. Springer-Verlag, Berlin.

Eshelman, L. and Schaffer, D. (1993). Real-coded genetic algorithms and interval-schemata. In Whitley, D., editor, *Foundations of Genetic Algorithms*, pages 187–202. Morgan Kaufmann, San Francisco (CA).

Essinger, J. (2014). *Ada's Algorithm: How Lord Byron's Daughter Ada Lovelace Launched the Digital Age*. Melville House, London.

Euler, L. (1736). Solutio problematis ad geometriam situs pertinentis. *Academiae scientiarum Petropolitanae*, 8: 128–140.

Evans, J. S. (2008). Dual-processing accounts of reasoning, judgment, and social cognition. *Annual Review of Psychology*, 59: 255–278.

Evans, J. S. and Frankish, K., editors (2009). *In Two Minds: Dual Processes and Beyond*. Oxford University Press, Oxford (UK).

Evans, J. S. and Stanovich, K. E. (2013). Dual-process theories of higher cognition: Advancing the debate. *Perspectives on Psychological Science*, 8: 223–241.

Even, S. (1979). *Graph Algorithms*. Computer Science Press, Rockville (MD).

Everingham, M., Gool, L. V., Williams, C. K. I., Winn, J. and Zisserman, A. (2010). The Pascal visual object classes (VOC) challenge. *Journal of Computer Vision*, 88: 303–338.

Farley, B. and Clark, W. A. (1954). Simulation of self-organizing systems by digital computer. *Transactions of the IRE Professional Group on Information Theory*, 4: 76–84.

Feest, U. (2005). Operationism in psychology: What the debate is about, what the debate should be about. *Journal of the History of the Behavioral Sciences*, 41: 131–149.

Feigenbaum, E. and McCorduck, P. (1983). *The Fifth Generation: Artificial Intelligence and Japan's Computer Challenge to the World*. Addison Wesley, Reading (MA).

Felleman, D. J. and Van Essen, D. C. (1991). Distributed hierarchical processing in the primate cerebral cortex. *Cerebral Cortex*, 1: 1–47.

Fenn, J. (2007). Understanding Gartner's hype cycles. Technical Report G00144727, Gartner Inc., Stamford (CT).

Ferris, J. (2020). *Behind the Enigma: The Authorised History of GCHQ, Britain's Secret Cyber-Intelligence Agency*. Bloomsbury Publishing, London.

Flack, J. C. (2018). Coarse-graining as a downward causation mechanism. *Philosophical Transactions of the Royal Society A*, 375: 20160338.

Flores, F. (2000). Heideggerian thinking and the transformation of business practice. In Wrathall, M. and Malpas, J., editors, *Heidegger, Coping, and Cognitive Science: Essays in Honor of Hubert L. Dreyfus*, pages 271–291. MIT Press, Cambridge (MA). Volume 2.

Fodor, J. (1968a). The appeal to tacit knowledge in psychological explanation. *The Journal of Philosophy*, 65: 627–640.

Fodor, J. (1968b). *Psychological Explanation: An Introduction to the Philosophy of Psychology*. Random House, New York.

Fodor, J. (1974). Special sciences (or: The disunity of science as a working hypothesis). *Synthese*, 28: 97–115.

Fodor, J. (1980). Methodological solipsism considered as a research strategy in cognitive psychology. *Behavioral and Brain Science*, 3: 63–73.

Fodor, J. (1983). *Modularity of Mind: An Essay on Faculty Psychology*. MIT Press, Cambridge (MA).

Fogel, L. J., Owens, A. J., and Walsh, M. J. (1966). *Artificial Intelligence Through Simulated Evolution*. John Wiley, New York.

Foot, P. (1967). The problem of abortion and the doctrine of the double effect. *Oxford Review*, 5: 5–15.

Ford, M. (2018). *Architects of Intelligence*. Packt Publishing, Birmingham (UK).

Francis, W. N. and Kucera, H. (1967). *Computational Analysis of Present-Day American English*. Brown University Press, Providence (RI).

Frege, G. (1879). *Begriffsschrift, eine der arithmetischen nachgebildete Formelsprache des reinen Denkens*. Louis Nebert, Halle a. S.

Frege, G. (1884). *Die Grundlagen der Arithmetik: Eine logisch-mathematische Untersuchung über den Begriff der Zahl*. W. Koebner, Breslau. Reprinted by Olms, Hildesheim, 1961.

French, R. M. (1990). Subcognition and the limits of the Turing test. *Mind*, 99: 53–65.

French, R. M. (2000a). The Chinese Room: Just Say "No". In *Proceedings of the Annual Meeting of the Cognitive Science Society*.

French, R. M. (2000b). The Turing test: The first 50 years. *Trends in Cognitive Sciences*, 4: 115–122.

Friston, K. (2009). The free-energy principle: A rough guide to the brain? *Trends in Cognitive Sciences*, 13: 293–301.

Friston, K. (2010). The free-energy principle: A unified brain theory? *Nature Reviews Neuroscience*, 11: 127–138.

Friston, K. (2012). A free energy principle for biological systems. *Entropy*, 14: 2100–2121.

Friston, K. and Kiebel, S. (2009). Predictive coding under the free–energy principle. *Philosophical transactions of the Royal Society B*, 364: 1211–1221.

Friston, K. and Stephan, K. E. (2007). Free–energy and the brain. *Synthese*, 159: 417–458.

Fukushima, K. (1980). Neocognitron: A self-organizing neural network model for a mechanism of pattern recognition unaffected by shift in position. *Biological Cybernetics*, 36: 193–202.

Fukushima, K. (1988). Neocognitron: A hierarchical neural network capable of visual pattern recognition. *Neural Networks*, 1: 119–130.

Gagne, J. and Andersen, M. (2012). A generative facade design method based on daylighting performance goals. *Journal of Building Performance Simulation*, 5: 141–154.

Gallagher, S. (2008). Are minimal representations still representations? *International Journal of Philosophical Studies*, 16: 351–369.

Gallagher, S. (2017). *Enactivist Interventions: Rethinking the Mind*. Oxford University Press, Oxford (UK).

Gandy, R. (1996). Human versus mechanical intelligence. In Millican, P. and Clark, A., editors, *Machines and Thought: The Legacy of Alan Turing*, volume I, pages 125–136. Oxford University Press, Oxford (UK).

Gandy, R. and Yates, M., editors (2001). *Mathematical Logic – Collected Works of A.M. Turing*. Elsevier, Amsterdam.

Garagnani, M., Wennekers, T. and Pulvermüller, F. (2008). A neuroanatomically-grounded Hebbian learning model of attention-language interactions in the human brain. *European Journal of Neuroscience*, 27: 492–513.

Gardner, H. (1985). *The Mind's New Science – A History of the Cognitive Revolution*. Basic Books, New York.

Gauss, C. F. (1900). *Werke*. Teubner, Leipzig.

Gauthier, I. and Tarr, M. J. (2016). Visual object recognition: Do we (finally) know more now than we did? *Annual Review of Vision Science*, 2: 16.1–16.20.

Gaze, R. M. (1958). The representation of the retina on the optic lobe of the frog. *Quarterly Journal of Experimental Physiology*, 43: 209–214.

Geeraerts, D., editor (2006). *Cognitive Linguistics: Basic Readings*. Mouton de Gruyter, Berlin.

Genova, J. (1994). Turing's sexual guessing game. *Social Epistemology: A Journal of Knowledge, Culture and Policy*, 8: 313–326.

Gigerenzer, G. (2007). *Gut Feelings: The Intelligence of the Unconscious*. Viking Press, New York.

Gigerenzer, G., Hertwig, R. and Pachur, T., editors (2011). *Heuristics: The foundations of adaptive behavior*. Oxford University Press, Oxford (UK).

Girshick, R. (2015). Fast R-CNN. In *Proceedings of IEEE International Conference on Computer Vision and Pattern Recognition*, pages 1440–1448.

Gobbo, F. and Durnová, H. (2014). From universal to programming languages. In *Informal Proceedings of Computability in Europe*, pages 1–10.

Godfrey, J., Holliman, E. and McDaniel, J. (1992). SWITCHBOARD: Telephone speech corpus for research and development. In *International Conference on Acoustics, Speech and Signal Processing*, pages 517–520.

Goldberg, A. (2006). *Constructions At Work. The Nature Of Generalization In Language*. Oxford University Press, Oxford (UK).

Goldstine, H. H. (1972). *The Computer: From Pascal to von Neumann*. Princeton University Press, Princeton (NJ).

Golovanov, S., Tselousov, A., Kurbanov, R. and Nikolenko, S. I. (2020). Lost in conversation: A conversational agent based on the Transformer and transfer learning. In Escalera and Herbrich (2020), pages 295–315.

Good, I. J. (1965). Speculations concerning the first ultraintelligent machine. In Alt, F. L. and Rubinoff, M., editors, *Advances in Computers*, volume 6, pages 31–88. Academic Press, New York.

Grantham, T. A. and Nichols, S. (1999). Evolutionary psychology: Ultimate explanations and Panglossian predictions. In Hardcastle, V. G., editor, *Philosophy and Linguistics*, pages 47–66. MIT Press, Cambridge (MA).

Grasso, G., Lucifora, C., Perconti, P. and Plebe, A. (2019). Evaluating mentalization during driving. In Gusikhin, O. and Helfert, M., editors, *5th International Conference on Vehicle Technology and Intelligent Transport Systems*, pages 536–541.

Green, C. D. (2005). Was Babbage's analytical engine intended to be a mechanical model of the mind? *History of Psychology*, 8: 35–45.

Grene, M. (1948). *Dreadful Freedom: A Critique of Existentialism*. Chicago University Press, Chicago (IL).

Grene, M. (1976). *Philosophy In and Out of Europe*. California University Press, Berkeley (CA).

Grene, M., editor (1983). *Dimensions of Darwinism: Themes and Counterthemes in Twentieth Century Evolutionary Theory*. Cambridge University Press, Cambridge (UK).

Grier, D. A. (2005). *When Computers Were Human*. Princeton University Press, Princeton (NJ).

Griffiths, P. E. (1996). The historical turn in the study of adaptation. *British Journal for the Philosophy of Science*, 47: 511–532.

Grill-Spector, K., Weiner, K. S., Gomez, J., Stigliani, A. and Natu, V. S. (2018). The functional neuroanatomy of face perception: From brain measurements to deep neural networks. *Interface Focus*, 8: 20180013.

Grosz, B. J., Sparck-Jones, K. and Webber, B. L. (1986). *Readings in Natural Language Processing*. Morgan Kaufmann, San Francisco (CA).

Güçlü, U. and van Gerven, M. A. J. (2014). Unsupervised feature learning improves prediction of human brain activity in response to natural images. *PLoS Computational Biology*, 10: 1–16.

Güçlü, U. and van Gerven, M. A. J. (2015). Deep neural networks reveal a gradient in the complexity of neural representations across the ventral stream. *Journal of Neuroscience*, 35: 10005–10014.

Gunderson, K. (1964). The imitation game. *Mind*, 73: 234–245.

Hacking, I. (1975). *The Emergence of Probability: A Philosophical Study of Early Ideas about Probability, Induction and Statistical Inference*. Cambridge University Press, Cambridge (UK).

Hacking, I. (1990). *The Taming of Chance*. Cambridge University Press, Cambridge (UK).

Hain, T., Woodland, P. C., Evermann, G., Gales, M. J. F., Liu, X., Moore, G. L., Povey, D. and Wang, L. (2005). Automatic transcription of conversational telephone speech. *IEEE Transactions on Speech and Audio Processing*, 13: 1173–1185.

Haken, H. (1964a). A nonlinear theory of laser noise and coherence. I. *Zeitschrift für Physik*, 181: 96–124.

Haken, H. (1964b). Theory of coherence of laser light. *Physical Review Letters*, 13: 329–331.

Haken, H. (1965). A nonlinear theory of laser noise and coherence. II. *Zeitschrift für Physik*, 182: 346–359.

Haken, H. (1969). Exact stationary solution of a Fokker-Plank equation for multimode laser action including phase locking. *Zeitschrift für Physik*, 219: 246–268.

Haken, H. (1978). *Synergetics – An Introduction, Nonequilibrium Phase Transitions and Self-Organization in Physics, Chemistry and Biology*. Springer-Verlag, Berlin, second edition.

Haken, H. (1983). *Advanced Synergetics – Instability Hierarchies of Self-Organizing Systems and Devices*. Springer-Verlag, Berlin.

Haken, H. (1984). *The Science of Structure: Synergetics*. Van Nostrand Reinhold, New York.

Haken, H. (1991). *Synergetic Computers and Cognition – A Top-Down Approach to Neural Nets*. Springer-Verlag, Berlin.

Haken, H. (1996). *Principles of Brain Functioning – A Synergetic Approach to Brain Activity, Behavior and Cognition*. Springer-Verlag, Berlin.

Haken, H. (2002). *Brain Dynamics – An Introduction to Models and Simulations*. Springer-Verlag, Berlin.

Han, J., Kamber, M. and Pei, J. (2012). *Data Mining: Concepts and Techniques*. Morgan Kaufmann, San Francisco (CA). 3rd edition.

Hanson, D. F. and White, V. (2004). Converging the capabilities of EAP artificial muscles and the requirements of bio-inspired robotics. In *Smart Structures and Materials*, pages 29–40.

Harary, F. (1969). *Graph Theory*. Addison Wesley, Reading (MA).

Harmon, L. D. (1959). Artificial neuron. *Science*, 129: 962–963.

Harnad, S. (1989). Minds, machines and Searle. *Journal of Experimental and Theoretical Artificial Intelligence*, 1: 5–25.

Harnad, S. (2000). Minds, machines and Turing – The indistinguishability of indistinguishables. *Journal of Logic, Language, and Information*, 9: 425–445.

Harris, R. A. (1993). *The Linguistics Wars*. Oxford University Press, Oxford (UK).

Hartline, H. K. (1967). Visual receptors and retinal interaction. *Science*, 164: 270–278.

Hassanien, A.-E., Dey, N. and Elghamrawy, S., editors (2020). *Big Data Analytics and Artificial Intelligence Against COVID-19: Innovation Vision and Approach*. Springer-Verlag, Berlin.

Haugeland, J. (1985). *Artificial Intelligence: The Very Idea*. MIT Press, Cambridge (MA).

Hauser, L. (2001). Look who's moving the goal posts now. *Minds and Machines*, 11: 41–51.

Hazelwood, K., Bird, S., Brooks, D., Chintala, S., Diril, U., Dzhulgakov, D., Fawzy, M., Jia, B., Jia, Y., Kalro, A., Law, J., Lee, K., Lu, J., Noordhuis, P., Smelyanskiy, M., Xiong, L. and Wang, X. (2018). Applied machine learning at Facebook: A datacenter infrastructure perspective. In *IEEE International Symposium on High Performance Computer Architecture (HPCA)*, pages 620–629.

Hebb, D. O. (1949). *The Organization of Behavior*. John Wiley, New York.

Heidegger, M. (1927). *Sein und Zeit*. Max Niemeyer Verlag, Tübingen, (DE). English Trans. *Being and Time* by John Macquarrie and Edward Robinson, 1962.

Heims, S. J. (1991). *The Cybernetics Group*. MIT Press, Cambridge (MA).

Hemlin, S. (1996). Research on research evaluation. *Social Epistemology*, 10: 209–250.

Hernàndez-Orallo, J. (2016). Evaluation in artificial intelligence: From task-oriented to ability-oriented measurement. *Artificial Intelligence Review*, 48: 397–447.

Herrera, C. and Sanz, R. (2016). Heideggerian AI and the being of robots. In Müller, V. C., editor, *Fundamental Issues of Artificial Intelligence*, pages 497–513. Springer-Verlag, Berlin.

Hilbert, D. and Ackermann, W. (1928). *Grundzüge der theoretischen Logik*. Springer-Verlag, Berlin. Trad. Principles of Mathematical Logic, Chelsea, New York, Second edition, 1959.

Hines, M. and Carnevale, N. (1997). The NEURON simulation environment. *Neural Computation*, 9: 1179–1209.

Hinton, G. E. (1981). Implementing semantic networks in parallel hardware. In Hinton, G. E. and Anderson, J. A., editors, *Parallel Models of Associative Memory*. Lawrence Erlbaum Associates, Mahwah (NJ).

Hinton, G. E., Osindero, S. and Teh, Y.-W. (2006). A fast learning algorithm for deep belief nets. *Neural Computation*, 18: 1527–1554.

Hinton, G. E. and Salakhutdinov, R. R. (2006). Reducing the dimensionality of data with neural networks. *Science*, 28: 504–507.

Hinton, G. E. and Sejnowski, T. J. (1983). Optimal perceptual inference. In *Proc. of IEEE International Conference on Computer Vision and Pattern Recognition*, pages 448–453, New York.

Hinton, G. E. and Sejnowski, T. J. (1986). Learning and relearning in Boltzmann machines. In Rumelhart and McClelland (1986b), pages 282–317.

Hinton, G. E., Sejnowski, T. J. and Ackley, D. H. (1984). Boltzmann machines: Constraint networks that learn. Technical Report 84-119, Carnegie-Mellon University, Computer Science Department.

Hinton, G. E., Srivastava, N., Krizhevsky, A., Sutskever, I. and Salakhutdinov, R. (2012). Improving neural networks by preventing co-adaptation of feature detectors. *arXiv*, abs/1207.0580.

Hirsch, J. E. (2005). An index to quantify an individual's scientific research output. *Proceedings of the Natural Academy of Science USA*, 46: 16569–16572.

Hobbes, T. (1651). *Leviathan*. London. English trans. by Edwin Curley, 1994, Indianapolis: Hackett.

Hobbes, T. (1655). *De Corpore*. London. English trans. by Aloysius Martinich, 1981, New York: Abaris Books.

Hochreiter, S. and Schmidhuber, J. (1997). Long short-term memory. *Neural Computation*, 9: 1735–1780.

Hodges, A. (1983). *Alan Turing: The Enigma*. Oxford University Press, Oxford (UK).

Hofstadter, D. R. (1981). Reflections. In Hofstadter, D. R. and Dennett, D. C., editors, *The Mind's I: Fantasies and Reflections on Self and Soul*. Basic Books, New York.

Holland, J. (1962). Outline for a logical theory of adaptive systems. *Journal of the ACM*, 9: 297–314.

Holland, J. (1975). *Adaptation in Natural and Artificial Systems*. University of Michigan Pres, Ann Arbor.

Holland, O. and Husbands, P. (2011). The origins of British cybernetics: The Ratio Club. *Kybernetes*, 40: 110–123.

Hopper, G. M. (1981). Keynote address. In Wexelblat, R., editor, *History of Programming Languages*. Academic Press, New York.

Hornik, K., Stinchcombe, M. and White, H. (1989). Multilayer feedforward networks are universal approximators. *Neural Networks*, 2: 359–366.

Hu, J., Shen, L. and Sun, G. (2018). Squeeze-and-excitation networks. In *Proceedings of IEEE International Conference on Computer Vision and Pattern Recognition*, pages 7132–7142.

Hubel, D. and Wiesel, T. (1959). Receptive fields of single neurones in the cat's striate cortex. *Journal of Physiology*, 148: 574–591.

Hubel, D. and Wiesel, T. (1962). Receptive fields, binocular interaction, and functional architecture in the cat's visual cortex. *Journal of Physiology*, 160: 106–154.

Hubel, D. and Wiesel, T. (1963). Shape and arrangement of columns in cat's striate cortex. *Journal of Physiology*, 165: 559–568.

Hubel, D. and Wiesel, T. (1968). Receptive fields and functional architecture of Mokey striate cortex. *Journal of Physiology*, 195: 215–243.

Hubel, D. and Wiesel, T. (1970). Stereoscopic vision in macaque monkey: Cells sensitive to binocular depth in area 18 of the macaque monkey cortex. *Nature*, 225: 41–42.

Hubel, D. and Wiesel, T. (1974). Uniformity of monkey striate cortex: A parallel relationship between field size, scatter, and magnification factor. *Journal of Comparative Neurology*, 158: 295–305.

Hume, D. (1739). *A Treatise of Human Nature*. John Noon, London. Vol. 1, 2.

Hutto, D. D. (1998). An ideal solution to the problems of consciousness. *Journal of Consciousness Studies*, 5: 328–343.

Hutto, D. D. (1999). Cognition without representation? In Riegler, A., Peschl, M., and von Stein, A., editors, *Understanding Representation in the Cognitive Sciences*, pages 57–74. Springer-Verlag, Berlin.

Hutto, D. D. (2013). Psychology unified: From folk psychology to radical enactivism. *Review of General Psychology*, 17: 174–178.

Hyneman, C. S. (1959). Means/ends analysis in policy science. *Political Research, Organization and Design*, 2: 19–22.

Hyönä, J., Tommola, J. and Alaja, A.-M. (1995). Pupil dilation as a measure of processing load in simultaneous interpretation and other language tasks. *Quarterly Journal of Experimental Psychology*, 48: 598–612.

Ichbiah, J. D., Krieg-Brueckner, B., Wichmann, B. A., Barnes, B. A. W. J. G. P., Roubine, J. G. P. B. O. M., Heliard, O. R. J. C. and Heliard, J.-C. (1979). Rationale for the design of the Ada programming language. *ACM Sigplan Notices*, 14: 1–261.

Ising, E. (1925). Beitrag zur Theorie des Rerromagnetismus. *Zeitschrift für Physik*, 31: 253–258.

Johansson, G. (1973). Visual perception of biological motion and a model for its analysis. *Perception and Psychophysics*, 14: 201–211.

Jones, C. and Spicer, A. (2009). *Unmasking the Entrepreneur*. Edward Elgar, Cheltenham (UK).

Jones, W., Alasoo, K., Fishman, D. and Parts, L. (2017). Computational biology: deep learning. *Emerging Topics in Life Sciences*, 1: 136–161.

Jong, K. D. (2005). Genetic algorithms: A 30-year perspective. In Booker, L., Forrest, S., Mitchell, M. and Riolo, R., editors, *Perspectives on Adaptation in Natural and Artificial Systems*, pages 11–31. Oxford University Press, Oxford (UK).

Jordan, M. I. (1986). Serial order: A Parallel Distributed Processing approach. Technical Report ICS-8604, University of California San Diego, Institute for Cognitive Science.

Kaas, J. H. (1997). Topographic maps are fundamental to sensory processing. *Brain Research Bulletin*, 44: 107–112.

Kadanoff, L. P. (2000). *Statistical Physics: Statics, Dynamics and Renormalization*. World Scientific Publishing, Singapore.

Kaplan, J. (2015). *Humans need not apply – A Guide to Wealth and Work in the Age of Artificial Intelligence*. Yale University Press, New Haven (CO).

Kasami, T. (1965). An efficient recognition and syntax algorithm for context-free languages. Scientific Report 65-758, AFCRL, Bedford (MA).

Katz, J. (1981). *Language and Other Abstract Objects*. Rowman and Littlefield, Totowa (NJ).

Katz, J. and Postal, P. (1991). Realism vs. conceptualism in linguistics. *Linguistics and Philosophy*, 14: 515–554.

Kay, K. N., Winawer, J., Mezer, A. and Wandell, B. A. (2013). Compressive spatial summation in human visual cortex. *Journal of Neurophysiology*, 110: 481–494.

Keil, F. C. (1989). *Concepts, Kinds, and Cognitive Development*. MIT Press, Cambridge (MA).

Keil, F. C. (1991). The emergence of theoretical beliefs as constraints on concepts. In Carey, S. and Gelman, R., editors, *The Epigenesis of Mind: Essays on Biology and Cognition*, pages 237–256. Cambridge University Press, Cambridge (UK).

Keil, F. C. (1994). The birth and nurturance of concepts by domains: The origins of concepts of living things. In Hirschfeld, L. A. and Gelman, S. A., editors, *Mapping the Mind: Domain Specificity in Cognition and Culture*, pages 234–254. Cambridge University Press, Cambridge (UK).

Ketkar, N. (2017). *Introduction to PyTorch*, pages 195–208. Apress, Berkeley (CA).

Khaligh-Razavi, S.-M. and Kriegeskorte, N. (2014). Deep supervised, but not unsupervised, models may explain IT cortical representation. *PLoS Computational Biology*, 10: e1003915.

Khan, S. and Tripp, B. P. (2017). One model to learn them all. *arXiv*, abs/1706.05137.

Kim, J. (1992). Multiple realization and the metaphysics of reduction. *Philosophy and Phenomenological Research*, 52: 1–26.

Kim, J. J., Pinker, S., Prince, A. and Prasada, S. (1991). Why no mere mortal has ever flown out to center field. *Cognitive Science*, 15: 173–218.

Kingma, D. P. and Ba, J. (2014). Adam: A method for stochastic optimization. In *Proceedings of International Conference on Learning Representations*.

Kingma, D. P. and Welling, M. (2014). Auto-encoding variational Bayes. In *Proceedings of International Conference on Learning Representations*.

Kisi, O. (2012). Forecasting daily lake levels using artificial intelligence approaches. *Computers & Geosciences*, 41: 169–180.

Kitaev, N. and Klein, D. (2018). Constituency parsing with a self-attentive encoder. *arXiv*, abs/1805.01052.

Kleene, S. C. (1952). *Introduction to Metamathematics*. Van Nostrand Reinhold, New York.

Klein, B., Harvey, B. M. and Dumoulin, S. O. (2014). Attraction of position preference by spatial attention throughout human visual cortex. *Neuron*, 84: 227–237.

Klemke, E. D., editor (1968). *Essays on Frege*. University of Illinois Press, Urbana (IL).

Kohonen, T. (1977). *Associative Memory – A System-Theoretical Approach*. Springer-Verlag, Berlin.

Kohonen, T. (1984). *Self-Organization and Associative Memory*. Springer-Verlag, Berlin.

Kohonen, T. (1995). *Self-Organizing Maps*. Springer-Verlag, Berlin.

Kohonen, T. (1998). The self-organizing map, a possible model of brain maps. In Pribram, K. H., editor, *Brain and Values: Is A Biological Science of Values Possible?*, pages 207–236. Psychology Press, East Sussex (UK).

Kohonen, T., Kaski, S., Lagus, K., Salojarvi, J., Honkela, J., Paatero, V. and Saarela, A. (2000). Self organization of a massive document collection. *IEEE Transactions on Neural Networks*, 11: 574–585.

Koller, D. and Friedman, N. (2009). *Probabilistic Graphical Models: Principles and Techniques*. MIT Press, Cambridge (MA).

Korb, K. B. and Nicholson, A. E. (2011). *Bayesian Artificial Intelligence*. Chapman and Hall / CRC, London.

Kornhuber, H. and Deecke, L. (1965). Hirnpotentialänderungen bei Willkü rbewegungen und passiven Bewegungen des Menschen: Bereitschaftspotential und reafferente Potentiale. *Pflügers Archiv für die Gesamte Physiologie des Menschen und der Tiere*, 284: 1–17.

Kouh, M. and Poggio, T. (2008). A canonical neural circuit for cortical nonlinear operations. *Neural Computation*, 20: 1427–1451.

Kriegeskorte, N. (2009). Relating population-code representations between man, monkey, and computational models. *Frontiers in Neuroscience*, 3: 363–373.

Kriegeskorte, N., Mur, M. and Bandettini, P. (2008). Representational similarity analysis – connecting the branches of systems neuroscience. *Frontiers in Systems Neuroscience*, 2: 4.

Krizhevsky, A. and Hinton, G. (2009). Learning multiple layers of features from tiny images. Technical Report Vol. 1, No. 4, University of Toronto.

Krizhevsky, A., Sutskever, I. and Hinton, G. E. (2012). ImageNet classification with deep convolutional neural networks. In *Advances in Neural Information Processing Systems*, pages 1090–1098.

Kuczaj, S. A. (1977). The acquisition of regular and irregular past tense forms. *Journal of Verbal Learning and Verbal Behavior*, 16: 589–600.

Kuffler, S. W. (1953). Discharge patterns and functional organization of mammalian retina. *Journal of Neurophysiology*, 16: 37–68.

Kushner, H. J. and Clark, D. (1978). *Stochastic Approximation Methods for Constrained and Unconstrained Systems*. Springer-Verlag, Berlin.

Kushwaha, S., Bahl, S., Bagha, A. K., Parmar, K. S., Javaid, M., Haleem, A. and Singh, R. P. (2020). Significant applications of machine learning for COVID-19 pandemic. *Journal of Industrial Integration and Management*, 5: 10.1142/S2424862220500268.

Lagus, K., Kaski, S., Honkela, T. and Kohonen, T. (1999). WEBSOM for textual data mining. *Artificial Intelligence Review*, 13: 345–364.

Lai, G., Xie, Q., Liu, H., Yang, Y. and Hovy, E. (2017). RACE: Large-scale reading comprehension dataset from examinations. In *Conference on Empirical Methods in Natural Language Processing*, pages 796–805.

Lakoff, G. (1986). A principled exception to the coordinate structure constraint. In *Proceedings of the the Twenty-First Regional Meeting, Chicago Linguistic Society*, Chicago (IL). Chicago Linguistic Society.

Langacker, R. W. (1987). *Foundations of Cognitive Grammar*. Stanford University Press, Stanford (CA).

Lashley, K. S. (1951). The problem of serial order in behavior. In Jeffress, L. A., editor, *Cerebral Mechanisms in Behavior*, pages 112–136. Bobbs-Merrill, Oxford, (UK).

LeCun, Y., Bengio, Y. and Hinton, G. (2015). Deep learning. *Nature*, 521: 436–444.

LeCun, Y., Bottou, L., Bengio, Y. and Haffner, P. (1998). Gradient-based learning applied to document recognition. *Proc. of the IEEE*, 86: 2278–2324.

Lee, J. (2017). *Sex Robots: The Future of Desire*. Palgrave MacMillan, London.

Lehky, S. R. and Tanaka, K. (2016). Neural representation for object recognition in inferotemporal cortex. *Current Opinion in Neurobiology*, 37: 23–35.

Leibniz, G. W. (1684). *Scientia Generalis. Characteristica.* in C. I. Gerhardt (ed.) *Die philosophischen Schriften – Siebenter Band*, Weidmann, Berlin, 1890.

Leroi-Gourhan, A. (1964). *Le geste et la parole.* Albin Michel, Paris. English translation by Bostock Berger, A. in *Gesture and Speech*, MIT Press, 1993.

Levenberg, K. (1944). A method for solution of certain non-linear problems in least squares. *Quarterly of Applied Mathematics*, 2: 164–168.

Levesque, H. J., editor (2017). *Common Sense, the Turing Test, and the Quest for Real AI.* MIT Press, Cambridge (MA).

Li, P., Farkas, I. and MacWhinney, B. (2004). Early lexical development in a self-organizing neural network. *Neural Networks*, 17: 1345–1362.

Libet, B. (2004). *Mind Time: The Temporal Factor in Consciousness.* Harvard University Press, Cambridge (MA).

Lighthill, J. (1973). Artificial intelligence: A general survey. Technical report, Science Research Council, London.

Linde, Y., Buzo, A. and Gray, R. (1980). An algorithm for vector quantizer design. *IEEE Transactions on Communications*, 28: 84–95.

Lindsay, G. W. (2020). Attention in psychology, neuroscience, and machine learning. *Frontiers in Computational Neuroscience*, 14: article 29.

Little, D. (1991). *Varieties of Social Explanation: An Introduction to the Philosophy of Social Science.* Westview Press, Boulder (CO).

Littman, E., Meyering, A., Walter, J., Wengerek, T. and Ritter, H. (1992). Neural networks for robotics. In Schuster, K., editor, *Applications of Neural Networks.* VHC, Weinheim, Germany.

Liu, W., Wang, Z., Liu, X., Zeng, N., Liu, Y. and Alsaadi, F. E. (2017). A survey of deep neural network architectures and their applications. *Neurocomputing*, 234: 11–26.

Llull, R. (1260–1310). *Opera Omnia.* Moguntiae, Paris. Collected by Ivo Salzinger et al. (eds), 1721–1742, Mainz; English trans. by Anthony Bonner, 1985, *Selected Works of Ramon Llull*, Princeton University Press.

Locke, J. (1690). *An Essay Concerning Human Understanding.* Meridian Books, Cleveland.

López-Rubio, E. (2018). Computational functionalism for the deep learning era. *Minds and Machines*, 28: 667–688.

Lothaire, M. (1997). *Combinatorics on Words*. Cambridge University Press, Cambridge (UK). Second Edition.

Lothaire, M. (2002). *Algebraic Combinatorics on Words*. Cambridge University Press, Cambridge (UK).

Lovelace, A. A. (1843). Translator's notes to an article on Babbage's analytical engine. *Richard Taylor's Scientific Memories*, 3: 666–731.

Lu, Y. (2019). Artificial intelligence: A survey on evolution, models, applications and future trends. *Journal of Management Analytics*, https://doi.org/10.1080/23270012.2019.1570365:1–29.

Lucifora, C., Grasso, G. M., Perconti, P., and Plebe, A. (2021). Moral reasoning and automatic risk reaction during driving. *Cognition, Technology & Work*, https://doi.org/10.1007/s10111-021-00675-y.

Luhmann, N. (1984). *Soziale Systeme: Grundrißeiner allgemeinen Theorie*. Suhrkamp Verlag, Frankfurt. English translation in *Social Systems*, Stanford University Press, 1995.

Luo, Y. and Xu, X. (2021). Comparative study of deep learning models for analyzing online restaurant reviews in the era of the COVID-19 pandemic. *International Journal of Hospitality Management*, 94: 102849.

MacWhinney, B. (1994). The dinosaurs and the ring. In Corrigan, R., Iverson, G., and Lima, S., editors, *The Reality of Linguistics Rules*, pages 283–320. John Benjamins, Amsterdam.

MacWhinney, B. (1998). Models of the emergence of language. *Annual Review of Psychology*, 49: 199–227.

MacWhinney, B. and Leinbach, J. (1991). Implementations are not conceptualizations: Revising the verb learning model. *Cognition*, 29: 121–157.

Magistretti, P. and Allaman, I. (2016). Brain energy metabolism. In Pfaff, D. W. and Volkow, N. D., editors, *Neuroscience in the 21st Century: From Basic to Clinical*, pages 1879–1910. Springer-Verlag, Berlin. Second Edition.

Maiman, T. H. (1960). Stimulated optical radiation in ruby. *Nature*, 187: 493–494.

Mallakin, A. (2019). An integration of deep learning and neuroscience for machine consciousness. *Global Journal of Computer Science and Technology*, 19: 1–10.

Mandler, J. M. (1988). How to build a baby: On the development of an accessible representational system. *Cognitive Development*, 3: 113–136.

Mandler, J. M. (1992). How to build a baby II: Conceptual primitives. *Psychological Review*, 99: 587–604.

Mandler, J. M., Bauer, P. J. and McDonough, L. (1991). Separating the sheep from the goats: Differentiating global categories. *Cognitive Psychology*, 23: 263–298.

Manzotti, R. and Chella, A. (2020). Conscious machines: A possibility? If so, how? *Journal of Artificial Intelligence and Consciousness*, 7: 183–198.

Marchand, R. (1993). The designers go to the fair II: Norman Bel Geddes, the General Motors 'Futurama', and the visit to the factory transformed. *Design Issues*, 8: 22–40.

Marcus, G. (1995). The acquisition of the English past tense in children and multilayered connectionist networks. *Cognition*, 46: 271–279.

Marcus, M., Santorini, B. and Marcinkiewicz, M. A. (1993). Building a large annotated corpus of English: The Penn Treebank. Technical Report MS-CIS-93-87, University of Pennsylvania Department of Computer and Information Science.

Markov, N. T., Ercsey-Ravasz, M. M., Gomes, A. R. R., Lamy, C., Magrou, L., Vezoli, J., Misery, P., Falchier, A., Quilodran, R., Gariel, M. A., Sallet, J., Gamanut, R., Huissoud, C., Clavagnier, S., Giroud, P., Sappey-Marinier, D., Barone, P., Dehay, C., Toroczkai, Z., Knoblauch, K., Essen, D. C. V. and Kennedy, H. (2014). A weighted and directed interareal connectivity matrix for macaque cerebral cortex. *Cerebral Cortex*, 24: 17–36.

Marr, D. (1969). A theory of cerebellar cortex. *Journal of Physiology*, 202:437–470.

Marr, D. (1970). A theory for cerebral neocortex. *Proceedings of the Royal Society of London B*, 176: 161–234.

Marr, D. (1971). Simple memory: A theory for archicortex. *Philosophical transactions of the Royal Society B*, 262: 23–81.

Marr, D. (1975). Review: Approaches to biological information processing. *Science*, 190: 875–876.

Marr, D. (1982). *Vision: A Computational Investigation into the Human Representation and Processing of Visual Information*. W. H. Freeman, San Francisco (CA).

Marr, D. and Hildreth, E. (1980). Theory of edge detection. *Proceedings of the Royal Society of London*, B207: 187–217.

Marr, D. and Nishihara, H. K. (1978). Representation and recognition of the spatial organization of three-dimensional shapes. *Proceedings of the Royal Society of London*, 200: 269–294.

Marr, D. and Poggio, T. A. (1979). From understanding computation to understanding neural circuitry. *Neuroscience Research Program Bulletin*, 15: 470–488.

Marshall, J. C. and Halligan, P. W. (1988). Blindsight and insight in visuo-spatial neglect. *Nature*, 336: 766–767.

Martinotti, C. (1890). Beitrag zum Studium der Hirnrinde und dem Centralursprung der Nerven. *Internationale Monatsschrift für Anatomie und Physiologie*, 7: 69–90.

Maturana, H. R. and Varela, F. J. (1980). *Autopoiesis and Cognition: The Realization of the Living*. Springer-Verlag, Berlin.

Maynard Smith, J. and Szathmáry, E. (2000). *Origins of Life: From the Birth of Life to the Origin of Language*. Oxford University Press, Oxford (UK).

Mayr, E. (1963). *Animal Species and Evolution*. Harvard University Press, Cambridge (MA).

Mays, W. (1951). Can machines think? *Philosophy*, 27: 148–162.

McCarthy, J., Minsky, M. L., Rochester, N. and Shannon, C. E. (1955). A proposal for the Dartmouth summer research project on artificial intelligence, August 31, 1955. reprinted in *AI Magazine*, 27: 12–14.

McCarthy, J. and Shannon, C. (1956). Preface. In Shannon, C. and McCarthy, J., editors, *Automata Studies*, pages i–iv. Princeton University Press, Princeton (NJ).

McCorduck, P. (1979). *Machines Who Think: A Personal Inquiry into the History and Prospect of Artificial Intelligence*. Freeman, San Francisco.

McCormac, E. R. (1976). *Metaphor and Myth in Science and Religion*. Duke University Press, Durham (NC).

McGinn, C. (1991). *The Problem of Consciousness*. Basil Blackwell, Oxford (UK).

McClelland, J. L. (1988). Connectionist models and psychological evidence. *Journal of Memory and Language*, 27: 107–123.

McClelland, J. L. and Patterson, K. (2002). Rules or connections in past-tense inflections: What does the evidence rule out? *Trends in Cognitive Sciences*, 6: 465–472.

McClellandi, J. L. and Roger, T. T. (2003). The parallel distributed processing approach to semantic cognition. *Nature*, 4: 310–322.

Mehta, P. and Schwab, D. J. (2014). An exact mapping between the variational renormalization group and deep learning. *arXiv*, abs/1410.03831.

Michalewicz, Z. (1994). *Genetic Algorithms + Data Structures = Evolution Programs*. Springer-Verlag, Berlin.

Michie, D. (2002). The very early days. Talk to the seminar 'Artificial Intelligence Recollections of the Pioneers'.

Miikkulainen, R. (1993). *Subsymbolic Natural Language Processing: An Integrated Model of Scripts, Lexicon and Memory*. MIT Press, Cambridge (MA).

Miikkulainen, R., Bednar, J., Choe, Y. and Sirosh, J. (2005). *Computational Maps in the Visual Cortex*. Springer-Science, New York.

Miikkulainen, R. and Dyer, M. G. (1991). Natural language processing with modular PDP networks and distributed lexicon. *Cognitive Science*, 15: 343–399.

Mikolov, T., Sutskever, I., Chen, K., Corrado, G. and Dean, J. (2013). Distributed representations of words and phrases and their compositionality. In *Advances in Neural Information Processing Systems*, pages 3111–3119.

Miller, J. and Bower, J. M. (2013). Introduction: Origin and history of the CNS meetings. In Bower, J. M., editor, *20 Years of Computational Neuroscience*, pages 1–13. Springer-Verlag, Berlin.

Minsky, M. (1954). *Neural Nets and the Brain-model Problem*. PhD thesis, Princeton University.

Minsky, M. (1968). Machines are more than they seem. *Science Journal*, 4: 3.

Minsky, M. (1986). *The Society of Mind*. Simon and Schuster, New York.

Minsky, M. and Papert, S. (1969). *Perceptrons: An Introduction to Computational Geometry*. MIT Press, Cambridge (MA).

Minsky, M. and Papert, S. (1972). Artificial intelligence progress report. Technical Report Memo n. 252, Massachussetts Institute of Technology.

Mitchell, M. and Forrest, S. (1994). Genetic algorithms and artificial life. *Artificial Life*, 1: 267–289.

Mithen, S. (1996). *The Prehistory of the Mind: The Cognitive Origins of Art, Religion and Science*. Thames and Hudson, London.

Mnih, V., Kavukcuoglu, K., Silver, D., Rusu, A. A., Veness, J., Bellemare, M. G., Graves, A., Riedmiller, M., Fidjeland, A. K., Ostrovski, G., Petersen, S., Beattie, C., Sadik, A., Antonoglou, I., King, H., Kumaran, D., Wierstra, D., Legg, S. and Hassabis, D. (2015). Human-level control through deep reinforcement learning. *Nature*, 518: 529–533.

Mohamed, A., Dahl, G. E. and Hinton, G. (2011). Acoustic modeling using deep belief networks. *IEEE Transactions on Audio, Speech, and Language Processing*, 20: 14–22.

Moor, J. H. (1976). An analysis of the Turing test. *Philosophical Studies*, 30: 249–257.

Moor, J. H., editor (2003). *The Turing Test – The Elusive Standard of Artificial Intelligence*. MIT Press, Cambridge (MA).

Moravec, H. (1988). *Mind Children: The Future of Robot and Human Intelligence*. Harvard University Press, Cambridge (MA).

Morin, A. (2006). Levels of consciousness and self-awareness: A comparison and integration of various neurocognitive views. *Consciousness and Cognition*, 15: 358–371.

Morin, A. (2018). The self-reflective functions of inner speech. thirteen years later. In Langland-Hassan, P. and Vicente, A., editors, *Inner speech: New voices*, pages 276–298. Oxford University Press, Oxford (UK).

Morris, W. C. and Cottrell, G. W. (1999). The empirical acquisition of grammatical relations. In Hahn, M. and Stoness, S. C., editors, *Proceedings of the Twenty First Annual Conference of the Cognitive Science Society*, pages 438–443. Lawrence Erlbaum Associates.

Morse, M. (1921). Recurrent geodesics on a surface of negative curvature. *Transaction of American Mathematical Society*, 22: 84–100.

Mountcastle, V. (1957). Modality and topographic properties of single neurons in cat's somatic sensory cortex. *Journal of Neurophysiology*, 20: 408–434.

Mountcastle, V. (1978). An organizing principle for cerebral function: The unit model and the distributed system. In Edelman, G. and Mountcastle, V., editors, *The Mindful Brain*. MIT Press, Cambridge (MA).

Moyer, A. E. (1991). P. W. Bridgman's operational perspective on physics part II: Refinements, publication, and reception. *Studies in History and Philosophy of Science*, 22: 373–397.

Newcomb, R. W. (1994). Some historical perspectives on early pulse coded neural network circuits. In Zaghloul, M. E., Meador, J. L., and Newcomb, R. W., editors, *Silicon implementation of pulse coded neural networks*, pages 1–8. Springer-Verlag, Berlin.

Newell, A. (1980). Physical symbol systems. *Cognitive Science*, 4: 135–183.

Newell, A., Shaw, C. and Simon, H. A. (1957). Empirical explorations of the logic theory machine: A case study in heuristic. In *Western Joint Computer Conference Proceedings*, pages 218–230, New York. ACM.

Newell, A., Shaw, C. and Simon, H. A. (1959). Report on a general problem-solving program. Scientific Report P-1584, RAND Corporation, Santa Monica (CA).

Newell, A. and Simon, H. A. (1972). *Human Problem Solving*. Prentice Hall, Englewood Cliffs (NJ).

Newell, A. and Simon, H. A. (1976). Computer science as empirical enquiry: Symbols and search. *Communications of the Association for Computing Machinery*, 19: 113–126.

Niu, J., Tang, W., Xu, F., Zhou, X. and Song, Y. (2016). Global research on artificial intelligence from 19902014: Spatially-explicit bibliometric analysis. *International Journal of Geo-Information*, 5: 66.

Nofre, D., Priestley, M. and Alberts, G. (2014). When technology became language: The origins of the linguistic conception of computer programming, 19501960. *Technology and Culture*, 55: 40–75.

Novaes, C. D. (2012). *Formal Languages in Logic: A Philosophical and Cognitive Analysis*. Cambridge University Press, Cambridge (UK).

Nunn, C. (2016). More splodge than singularity? In Awret, U., editor, *The Singularity: Could Artificial Intelligence Really Out-Think Us (and Would We Want It To)?*, pages 408–412, New York. Imprint Academic.

Oakes, K. (2019). Here's what scientists searched for in 2018: AI is up, stress is down. *Nature*, doi: 10.1038/d41586-018-07879-9.

Ofner, A. and Stober, S. (2018). Towards bridging human and artificial cognition: Hybrid variational predictive coding of the physical world, the body and the brain. In *Advances in Neural Information Processing Systems*.

Okrent, M. (2003). Heidegger in America or how transcendental philosophy becomes pragmatic. In Malpas, J., editor, *From Kant to Davidson – Philosophy and the Idea of the Transcendental*, pages 122–138. Routledge, London.

Olazaran, M. (1996). A sociological study of the official history of the perceptrons controversy. *Social Studies of Science*, 26: 611–659.

O'Leary, D. (1997). The internet, intranets, and the AI renaissance. *Computer*, 30: 71–78.

O'Leary, D. (2001). Thanks for the renaissance! *IEEE Intelligent Systems*, 16: 2–3.

Olshausen, B. A. (2014). Perception as an inference problem. In Gazzaniga, M. S., editor, *The Cognitive Neurosciences*, pages 295–304. MIT Press, Cambridge (MA). Fifth edition.

Osterhout, L., Kim, A. and Kuperberg, G. R. (2007). The neurobiology of sentence comprehension. In Spivey, M., McRae, K. and Joanisse, M., editors, *The Cambridge Handbook of Psycholinguistics*, pages 365–389. Cambridge University Press, Cambridge (UK).

Panksepp, J. and Panksepp, J. B. (2000). The seven sins of evolutionary psychology. *Evolution and Cognition*, 6: 108–131.

Papert, S. (1966). The summer vision project. Technical Report Vision Memo n. 100, Massachussetts Institute of Technology.

Parisi, G. (1988). *Statistical Field Theory*. Addison Wesley, Reading (MA).

Patera, V. (2004). Alan's Apple: Hacking the Turing Test. In Teuscher (2004), pages 9–41.

Pearl, J. (1986). Fusion, propagation, and structuring in belief networks. *Artificial Intelligence*, 29: 241–288.

Pearl, J. (1988). *Probabilistic Reasoning in Intelligent Systems: Networks of Plausible Inference*. Morgan Kaufmann, San Francisco (CA).

Pearl, J. and Mackenzie, D. (2018). *The Book of Why – The New Science of Cause and Effect*. Hachette Book Group, New York.

Pennington, J., Socher, R. and Manning, C. D. (2014). GloVe: Global vectors for word representation. In *Conference on Empirical Methods in Natural Language Processing*, pages 1532–1543. Association for Computational Linguistics.

Pepperell, R. (2018). Consciousness as a physical process caused by the organization of energy in the brain. *Frontiers in Psychology*, 9: 2091.

Perconti, P. (1999). *Kantian Linguistics: Theories of Mental Representation and the Linguistic Transformation in Kantism*. Nodus Publikationen, Münster (DE).

Perconti, P. (2002). Context-dependence in human and animal communication. *Foundations of Science*, 7: 341–362.

Perconti, P. (2020). Moderate mindreading priority. In Pennisi, A. and Falzone, A., editors, *The Extended Theory of Cognitive Creativity – Interdisciplinary Approaches to Performativity*, pages 103–113. Springer-Verlag, Berlin.

Perconti, P. and Plebe, A. (2020). Deep learning and cognitive science. *Cognition*, 203: Article 104365.

Perconti, P. and Zeppi, A. (2014). Mindreading and computational tractability. *Anthropology & Philosophy*, 11: 1.

Perrault, R., Shoham, Y., Brynjolfsson, E., Clark, J., Etchemendy, J., Grosz, B., Lyons, T., Manyika, J. and Niebles, J. C. (2019). The AI index 2019 annual report. Technical report, Human-Centered AI Initiative, Stanford University, Stanford (CA).

Pham, H., Guan, M. Y., Zoph, B., Le, Q. V. and Dean, J. (2018). Efficient neural architecture search via parameter sharing. *arXiv*, abs/1802.03268.

Piccinini, G. (2009). Computationalism in the philosophy of mind. *Philosophy Compass*, 4: 515–532.

Pierson, H. A. and Gashler, M. S. (2017). Deep learning in robotics: A review of recent research. *Advanced Robotics*, 31: 821–835.

Pinker, S. (1991). Rules of language. *Science*, 253: 530–535.

Pinker, S. (1994). *The Language Instinct. How the Mind Creates Language*. William Morrow, New York.

Pinker, S. (1997). *How the Mind Works*. W. W. Norton & Company, New York.

Pinker, S. (1999). *Words and Rules: The Ingredients of Language*. Basic Books, New York.

Pinker, S. (2001). Four decades of rules and associations, or whatever happened to the past tense debate? In Dupoux, E., editor, *Language, Brain, and Cognitive Development: Essays in Honor of Jacques Mehler*, pages 157–179. MIT Press, Cambridge (MA).

Pinker, S. (2002). *The Blank Slate: The Modern Denial of Human Nature*. Penguin Books, New York.

Pinker, S. and Prince, A. (1988). On language and connectionism: Analysis of a parallel distributed processing model of language acquisition. *Cognition*, 28: 73–193.

Pinker, S. and Ullman, M. T. (2002a). Combination and structure, not gradedness, is the issue. *Trends in Cognitive Sciences*, 6: 472–475.

Pinker, S. and Ullman, M. T. (2002b). The past and future of the past tense. *Trends in Cognitive Sciences*, 6: 456–1338.

Pinsky, L. (1951). Do machines think about machines thinking? *Mind*, 60: 397–398.

Plebe, A. (2001). Self-organizing map approaches to the traveling salesman problem. In Maggini, M., editor, *Limitations and Future Trends in Neural Computation, NATO Advanced Research Workshop, 22–24 October, 2001, Siena, Italy.*

Plebe, A. (2018). The search of "canonical" explanations for the cerebral cortex. *History and Philosophy of the Life Sciences*, 40: 40–76.

Plebe, A. and Anile, M. (2001). A neural-network-based approach to the double traveling salesman problem. *Neural Computation*, 14(2): 437–471.

Plebe, A. and De La Cruz, V. M. (2016). *Neurosemantics – Neural Processes and the Construction of Linguistic Meaning*. Springer, Berlin.

Plebe, A. and Domenella, R. G. (2007). Object recognition by artificial cortical maps. *Neural Networks*, 20: 763–780.

Plebe, A. and Grasso, G. (2019). The unbearable shallow understanding of deep learning. *Minds and Machines*, 29: 515–553.

Plebe, A. and Perconti, P. (2013). The slowdown hypothesis. In Eden, A. H., Moor, J. H., Søraker, J. H., and Steinhart, E., editors, *Singularity Hypotheses*. Springer, Berlin.

Plebe, A. and Perconti, P. (2020). Plurality: The end of singularity? In Korotayev, A. V. and LePoire, D. J., editors, *The 21st Century Singularity and Global Futures – A Big History Perspective*, pages 163–184. Springer, Berlin.

Plunkett, K. and Juola, P. (1999). A connectionist model of English past tense and plural morphology. *Cognition*, 48: 21–69.

Plunkett, K. and Marchman, V. A. (1993). From rote learning to system building: Acquiring verb morphology in children and connectionist nets. *Cognition*, 48: 21–69.

Plunkett, K. and Marchman, V. A. (1996). Learning from a connectionist model of the acquisition of the English past tense. *Cognition*, 61: 299–308.

Polak, E. (1971). *Computational Methods in Optimization: A Unified Approach.* Academic Press, New York.

Popa, R. (2004). *Between Necessity and Probability: Searching for the Definition and Origin of Life.* Springer-Verlag, Berlin.

Post, E. L. (1936). Finite combinatory processes – Formulation 1. *The Journal of Symbolic Logic*, 1: 103–105.

Post, E. L. (1947). Recursive unsolvability of a problem of Thue. *Journal of Symbolic Logic*, 12: 1–11.

Prasada, S. and Pinker, S. (1993). Generalisation of regular and irregular morphological patterns. *Language and Cognitive Processes*, 8:1–56.

Preston, J. and Bishop, M., editors (2002). *Views into the Chinese Room: New Essays on Searle and Artificial Intelligence.* Oxford University Press, Oxford (UK).

Proverbio, A. M., Crotti, N., Zani, A. and Adorni, R. (2009). The role of left and right hemispheres in the comprehension of idiomatic language: An electrical neuroimaging study. *BMC Neuroscience*, 10: 116.

Putnam, H. (1960). Minds and machines. In Hook, S., editor, *Dimensions of Mind*, pages 138–164. New York University Press, New York.

Putnam, H. (1967a). The mental life of some machines. In Castañeda, H. - N., editor, *Intentionality, Minds and Perception*, pages 177–200. Wayne State University Press, Detroit (MI).

Putnam, H. (1967b). Psychological predicates. In Capitan, W. and Merrill, D., editors, *Art, Mind, and Religion*, pages 37–48. University of Pittsburgh Press, Pittsburgh (PA).

Putnam, H. (1975). Philosophy and our mental life. In Putnam, H., editor, *Mind, Language and Reality*, volume 2, pages 291–303. MIT Press, Cambridge (MA).

Pylyshyn, Z. (1981). Computation and cognition: Issues in the foundations of cognitive science. *Behavioral and Brain Science*, 3: 111–150.

Pylyshyn, Z. (1984). *Computation and Cognition.* MIT Press, Cambridge (MA).

Quillian, M. R. (1967). Word concepts: A theory and simulation of some basic semantic capabilities. *Behavioral Science*, 12: 410–430.

Quillian, M. R. (1968). Semantic memory. In Minsky, M., editor, *Semantic Information Processing*, pages 227–270. MIT Press, Cambridge (MA).

Quine, W. V. O. (1955). On Frege's way out. *Mind*, 64:145–159. Reprinted in Klemke (1968).

Rajaraman, A. and Ullman, J. D. (2011). *Mining of Massive Datasets*. Cambridge University Press, Cambridge (UK).

Ramsey, F. P. (1931). Truth and probability. In Braithwaite, R., editor, *The Foundations of Mathematics and other Logical Essays*, pages 156–198. The Humanities Press, New York.

Randell, B., editor (1973). *The Origins of Digital Computers*. Springer-Verlag, Berlin.

Rapaport, W. J. (2000). How to pass a Turing test – Syntactic semantics, natural-language understanding, and first-person cognition. *Journal of Logic, Language, and Information*, 9: 467–490.

Rawat, W. and Wang, Z. (2017). Deep convolutional neural networks for image classification: A comprehensive review. *Neural Computation*, 29: 2352–2449.

Rechenberg, I. (1973). *Evolutionsstrategie – Optimierung technischer Systeme nach Prinzipien der biologischen Evolution*. Frommann-Holzboog, Stuttgard (DE).

Regier, T. (1995). A model of the human capacity for categorizing spatial relations. *Cognitive Linguistics*, 6: 63–88.

Reichardt, W. and Poggio, T. (1976). Visual control of orientation behaviour in the fly: Part I. A quantitative analysis. *Quarterly Reviews of Biophysics*, 9: 311–375.

Reichenbach, H. (1949). *The Theory of Probability*. Chicago University Press, Chicago (IL).

Retherford, J. (1975). Applications of Banach ideals of operators. *Bulletin of the American Mathematical Society*, 81: 978–1012.

Rezende, D. J., Mohamed, S. and Wierstra, D. (2014). Stochastic backpropagation and approximate inference in deep generative models. In Xing, E. P. and Jebara, T., editors, *Proceedings of Machine Learning Research*, pages 1278–1286.

Richardson, K. (2015). *An Anthropology of Robots and AI: Annihilation Anxiety and Machines*. Routledge, London.

Richardson, K. (2018). *Challenging Sociality: An Anthropology of Robots, Autism, and Attachment*. Palgrave MacMillan, London.

Richardson, M., Burges, C. J. and Renshaw, E. (2013). MCTest: A challenge dataset for the open-domain machine comprehension of text. In *Conference on Empirical Methods in Natural Language Processing*, pages 193–203.

Richardson, R. (2007). *Evolutionary Psychology as Maladapted Psychology*. MIT Press, Cambridge (MA).

Riesenhuber, M. and Poggio, T. (1999). Hierarchical models of object recognition in cortex. *Nature Neuroscience*, 2: 1019–1025.

Righi, R., Samoili, S., Cobo, M. L., Baillet, M. V., Cardona, M. and De Prato, G. (2020). The AI techno-economic complex system: Worldwide landscape, thematic subdomains and technological collaborations. *Telecommunications Policy*, https://doi.org/10.1016/j.telpol.2020.101943. In Press.

Risken, H. (1989). *The Fokker-Planck Equation – Methods of Solution and Applications*. Springer-Verlag, Berlin.

Ritter, H. and Kohonen, T. (1989). Self-organizing semantic maps. *Biological Cybernetics*, 61: 241–254.

Ritter, H., Martinetz, T., and Schulten, K. (1992). *Neural Computation and Self-Organizing Maps*. Addison Wesley, Reading (MA).

Robbins, H. and Monro, S. (1951). A stochastic approximation method. *Annals of Mathematical Statistics*, 22: 400–407.

Robbins, L. (1932). *An Essay on the Nature and Significance of Economic Science*. Macmillan, London.

Robinson, L. and Rolls, E. T. (2015). Invariant visual object recognition: Biologically plausible approaches. *Biological Cybernetics*, 109: 505–535.

Rogers, C. (1942). *Counseling and Psychotherapy: Newer Concepts in Practice*. Houghton Mifflin, Boston (MA).

Rogers, T. T. and McClelland, J. L. (2006). *Semantic Cognition - A Parallel Distributed Processing Approach*. MIT Press, Cambridge (MA).

Rogers, T. T. and McClelland, J. L. (2014). Parallel distributed processing at 25: Further explorations in the microstructure of cognition. *Cognitive Science*, 38: 1024–1077.

Rolls, E. (2016). *Cerebral Cortex: Principles of Operation*. Oxford University Press, Oxford (UK).

Rolls, E. and Deco, G. (2002). *Computational Neuroscience of Vision*. Oxford University Press, Oxford (UK).

Rolls, E. T. and Stringer, S. M. (2006). Invariant visual object recognition: A model, with lighting invariance. *Journal of Physiology – Paris*, 100: 43–62.

Rolnick, D., Donti, P. L., Kaack, L. H., Kochanski, K., Lacoste, A., Sankaran, K., Ross, A. S., Milojevic-Dupont, N., Jaques, N., Waldman-Brown, A., Luccioni, A., Maharaj, T., Sherwin, E. D., Mukkavilli, S. K., Gomes, K. P. K. C., Ng, A. Y., Hassabis, D., Platt, J. C., Creutzig, F., Chayes, J. and Bengio, Y. (2019). Tackling climate change with machine learning. *arXiv*, abs/1906.05433.

Rose, N. and Abi-Rached, J. M. (2013). *Neuro: The New Brain Sciences and the Management of the Mind*. Princeton University Press, Princeton (NJ).

Rosen, R. (1958). A relational theory of biological systems. *Bulletin of Mathematical Biophysics*, 20: 245–260.

Rosen, R. (1991). *Life Itself: A Comprehensive Inquiry into the Nature, Origin, and Fabrication of Life*. Columbia University Press, New York.

Rosenblatt, F. (1958). The perceptron: A probabilistic model for information storage and organisation in the brain. *Psychological Review*, 65: 386–408.

Rosenblatt, F. (1959). Two theorems of statistical separability in the perceptron. In *Mechanisation of thought processes: Proceedings of symposium No. 10*, pages 419–472, London. H. M. Stationery Office.

Rosenblatt, F. (1962). *Principles of Neurodynamics: Perceptron and the Theory of Brain Mechanisms*. Spartan, Washington (DC).

Rosenfeld, A. and Kak, A. C. (1982). *Digital Picture Processing*. Academic Press, New York, Second Edition.

Ruhland, K., Peters, C. E., Andrist, S., Badler, J. B., Badler, N. I., Gleicher, M., Mutlu, B. and McDonnell, R. (2015). A review of eye gaze in virtual agents, social robotics and HCI: Behaviour generation, user interaction and perception. *Computer Graphics Forum*, 34: 299–326.

Rumelhart, D. E. (1990). Brain style computation: Learning and generalization. In Zornetzer, S. F., Davis, J. L. and Lau, C., editors, *An Introduction to Neural and Electronic Networks*, pages 405–420. Academic Press, New York.

Rumelhart, D. E., Durbin, R., Golden, R. and Chauvin, Y. (1995). Backpropagation: The basic theory. In Chauvin, Y. and Rumelhart, D. E., editors, *Backpropagation: Theory, Architectures and Applications*, pages 1–34. Lawrence Erlbaum Associates, Mahwah (NJ).

Rumelhart, D. E., Hinton, G. E. and Williams, R. J. (1986). Learning representations by back-propagating errors. *Nature*, 323: 533–536.

Rumelhart, D. E. and McClelland, J. L. (1986a). On learning the past tenses of English verbs. In Rumelhart and McClelland (1986b), pages 216–271.

Rumelhart, D. E. and McClelland, J. L., editors (1986b). *Parallel Distributed Processing: Explorations in the Microstructure of Cognition.* MIT Press, Cambridge (MA).

Rumelhart, D. E. and Todd, P. M. (1993). Learning and connectionist representations. In Meyer, D. E. and Kornblum, S., editors, *Attention and Performance XIV: Synergies in Experimental Psychology*, pages 405–420. Academic Press, New York.

Russakovsky, O., Deng, J., Su, H., Krause, J., Satheesh, S., Ma, S., Huang, Z., Karpathy, A., Khosla, A., Bernstein, M., Berg, A. C. and Fei-Fei, L. (2015). ImageNet large scale visual recognition challenge. *International Journal of Computer Vision*, 115: 211–252.

Russell, B. (1902). Letter to Frege. in van Heijenoort (1967).

Russell, B. (1918). *Mysticism and Logic and Other Essays.* George Allen & Unwin, London.

Sachan, M., Dubey, A., Xing, E. P. and Richardson, M. (2015). Learning answer-entailing structures for machine comprehension. In *Annual Meeting of the Association for Computational Linguistics*, pages 239–249.

Safran, I. and Shamir, O. (2017). Depth-width tradeoffs in approximating natural functions with neural networks. *arXiv*, abs/1610.09887.

Salakhutdinov, R. R. and Hinton, G. E. (2009). Deep Boltzmann machines. In *International Conference on Artificial Intelligence and Statistics*, pages 448–455.

Salsburg, D. (2001). *Lady Tasting Tea – How Statistics Revolutionized Science in the Twentieth Century.* W. H. Freeman and Company, New York.

Samuel, A. (1962). Artificial intelligence: A frontier of automation. *The Annals of the American Academy of Political and Social Science*, 340: 10–20.

Sánchez, J. and Perronnin, F. (2011). High-dimensional signature compression for large-scale image classification. In *Proceedings of IEEE International Conference on Computer Vision and Pattern Recognition*, pages 1665–1672.

Sandini, G. and Sciutti, A. (2018). Humane robots–From robots with a humanoid body to robots with an anthropomorphic mind. *ACM Transactions on Human-Robot Interaction*, 16: 7.

Saon, G., Kurata, G., Sercu, T., Audhkhasi, K., Thomas, S., Dimitriadis, D., Cui, X., Ramabhadran, B. et al. (2017). English conversational telephone speech recognition by humans and machines. In *Conference of the International Speech Communication Association*, pages 132–136.

Savage, L. J. (1954). *The Foundations of Statistics*. John Wiley, New York.

Saygin, A. P., Cicekli, I. and Akman, V. (2000). Turing test: 50 years later. *Minds and Machines*, 10: 463–518.

Schmitt, L. M. (2001). Theory of genetic algorithms. *Theoretical Computer Science*, 130: 600–620.

Schrödinger, E. (1944). *What is Life? The Physical Aspect of the Living Cell*. Cambridge University Press, Cambridge (UK).

Schützenberger, M.-P. (1955). Une théorie algébrique du codage. *Séminaire Dubreil. Algèbre et théorie des nombres*, 9: 1–24.

Searle, J. R. (1980). Mind, brain and programs. *Behavioral and Brain Science*, 3: 417–424.

Searle, J. R. (1992). *The Rediscovery of the Mind*. MIT Press, Cambridge (MA).

Sejnowski, T. J., Koch, C. and Churchland, P. S. (1988). Computational neuroscience. *Science*, 241: 1299.

Seuren, P. (2004). *Chomsky's Minimalism*. Oxford University Press, Oxford (UK).

Shannon, C. (1956). A universal turing machine with two internal states. In Shannon, C. and McCarthy, J., editors, *Automata Studies*, pages 157–165. Princeton University Press, Princeton (NJ).

Shepherd, G. M. (1988). A basic circuit for cortical organization. In Gazzaniga, M. S., editor, *Perspectives on Memory Research*, pages 93–134. MIT Press, Cambridge (MA).

Shieber, S. M., editor (2004). *The Turing Test: Verbal Behavior as the Hallmark of Intelligence*. MIT Press, Cambridge (MA).

Shindo, H., Miyao, Y., Fujino, A. and Nagata, M. (2012). Bayesian symbol-refined tree substitution grammars for syntactic parsing. In *50th Annual Meeting of the Association for Computational Linguistics*, pages 440–448, Somerset (NJ). Association for Computational Linguistics.

Shoham, Y., Perrault, R., Brynjolfsson, E., Clark, J., Manyika, J., Niebles, J. C., Lyons, T., Etchemendy, J., Grosz, B. and Bauer, Z. (2018). The AI index 2018 annual report. Technical report, Human-Centered AI Initiative, Stanford University, Stanford (CA).

Siegmund-Schultze, R. (1997). The emancipation of mathematical research publishing in the United States from German dominance (1878–1945). *Historia Mathematica*, 24: 135–166.

Silver, D., Huang, A., Maddison, C. J., Guez, A., Sifre, L., van den Driessche, G., Schrittwieser, J., Antonoglou, I., Panneershelvam, V., Lanctot, M., Dieleman, S., Grewe, D., Nham, J., Kalchbrenner, N., Sutskever, I., Lillicrap, T., Leach, M., Kavukcuoglu, K., Graepel, T. and Hassabis, D. (2016). Mastering the game of Go with deep neural networks and tree search. *Nature*, 529: 484–489.

Simion, F., Regolin, L. and Bulf, H. (2008). A predisposition for biological motion in the newborn baby. *Proceedings of the Natural Academy of Science USA*, 105: 809–813.

Simon, H. A. (1947). *Administrative Behavior: A Study of Decision-Making Processes in Administrative Organization*. MacMillan, New York.

Simon, H. A. (1957). *Models of Man, Social and Rational: Mathematical Essays on Rational Human Behaviour in a Social Setting*. John Wiley, New York.

Simon, H. A. (1977). *Models of Discovery*. Reidel Publishing Company, Dordrecht (NL).

Simon, H. A. (1991). *Models of my Life*. Basic Books, New York.

Simon, H. A. and Newell, A. (1958). Heuristic problem solving: The next advance in operations research. *Operations Research*, 6: 1–10.

Simon, J. C. (1986). *Patterns and Operators: The Foundations of Data Representations*. Mc Graw Hill, New York.

Simonyan, K. and Zisserman, A. (2015). Very deep convolutional networks for large-scale image recognition. *arXiv*, abs/1409.1556.

Sirosh, J. and Miikkulainen, R. (1997). Topographic receptive fields and patterned lateral interaction in a self-organizing model of the primary visual cortex. *Neural Computation*, 9: 577–594.

Sivanandam, S. N. and Deepa, S. N. (2008). *Introduction to Genetic Algorithms*. Springer, Berlin.

Skinner, B. F. (1957). *Verbal Behavior*. Appleton-Century-Croft, New York.

So, D. R., Liang, C. and Le, Q. V. (2019). The evolved Transformer. In *International Conference on Machine Learning*.

Soon, C. S., Brass, M., Heinze, H.-J. and Haynes, J.-D. (2008). Unconscious determinants of free decisions in the human brain. *Nature Neuroscience*, 11: 543–545.

Soon, C. S., He, A. H., Bode, S. and Haynes, J.-D. (2013). Predicting free choices for abstract intentions. *Proceedings of the Natural Academy of Science USA*, 110: 5733–5734.

Spinosa, C., Flores, F. and Dreyfus, H. (1997). *Disclosing New Worlds: Entrepreneurship, Democratic Action, and the Cultivation of Solidarity*. MIT Press, Cambridge (MA).

Stahlberg, F., Saunders, D., de Gispert, A. and Byrne, B. (2019). CUED@WMT19:EWC&LMs. *arXiv*, abs/1906.05447.

Stich, S. P. and Ravenscroft, I. (1994). What is folk psychology? *Cognition*, 50: 447–468.

Stigliani, A., Jeska, B. and Grill-Spector, K. (2017). Encoding model of temporal processing in human visual cortex. *Proceedings of the Natural Academy of Science USA*, 1914: E11047–E11056.

Stinchcombe, M. and White, H. (1989). Universal approximation using feedforward networks with non-sigmoid hidden layer activation functions. In *Proceedings International Joint Conference on Neural Networks*, pages 613–617, S. Diego (CA).

Stowe, L. A., Haverkort, M. and Zwarts, F. (2004). Rethinking the neurological basis of language. *Lingua*, 115: 997–1042.

Stringer, S. M. and Rolls, E. T. (2002). Invariant object recognition in the visual system with novel views of 3D objects. *Neural Computation*, 14: 2585–2596.

Stringer, S. M., Rolls, E. T. and Tromans, J. M. (2007). Invariant object recognition with trace learning and multiple stimuli present during training. *Network: Computation in Neural Systems*, 18: 161–187.

Stueckelberg, E. and Petermann, A. (1953). La normalisation des constantes dans la théorie des quanta. *Helvetica Physica Acta*, 26: 499–520.

Sun, S., Chen, W., Wang, L., Liu, X. and Liu, T.-Y. (2016). On the depth of deep neural networks: A theoretical view. In *AAAI Conference on Artificial Intelligence*, pages 2066–2072.

Sünderhauf, N., Brock, O., Scheirer, W., Hadsell, R., Fox, D., Leitner, J., Upcroft, B., Abbeel, P., Burgard, W., Milford, M. and Corke, P. (2018). The limits and potentials of deep learning for robotics. *The International Journal of Robotics Research*, 37: 405–420.

Swade, D. D. (2001). *The Difference Engine: Charles Babbage*. Viking, New York.

Sylvester, J. J. (1878). A biographical sketch of a young infant. *Kosmos*, 1: 64–104.

Szegedy, C., Liu, W., Jia, Y., Sermanet, P., Reed, S., Anguelov, D., Erhan, D., Vanhoucke, V. and Rabinovich, A. (2015). Going deeper with convolutions. In *Proceedings of IEEE International Conference on Computer Vision and Pattern Recognition*, pages 1–9.

Tacchetti, A., Isik, L. and Poggio, T. A. (2018). Invariant recognition shapes neural representations of visual input. *Annual Review of Vision Science*, 4: 403–422.

Takeno, J., Inaba, K. and Suzuki, T. (2005). Experiments and examination of mirror image cognition using a small robot. In *Proceedings IEEE International Symposium on Computational Intelligence in Robotics and Automation*, pages 493–498.

Tan, K.-H. and Lim, B. P. (2018). The artificial intelligence renaissance: Deep learning and the road to human-level machine intelligence. *APSIPA Transactions on Signal and Information Processing*, 7: e6.

Teuscher, C., editor (2004). *Alan Turing: Life and Legacy of a Great Thinker*. Springer-Verlag, Berlin.

Thue, A. (1906). Über unendliche Zeichenreihen. *Norske Vid. Selsk. Skr. I. Mat. Nat. Kl.*, 7:1–22. English translation in T. Nagel et al. *Selected mathematical papers of Axel Thue*, Universitetsforlaget, Oslo, 1977.

Thue, A. (1912). Über die gegnseitige Lage gleiche Teile gewisser Zeichenreihen. *Norske Vid. Selsk. Skr. I. Mat. Nat. Kl.*, 10:1–67. English translation in T. Nagel et al. *Selected mathematical papers of Axel Thue*, Universitetsforlaget, Oslo, 1977.

Tolman, E. C. (1948). Cognitive maps in rats and men. *Psychological Review*, 55: 189–208.

Tomasello, M. (1999). *The Cultural Origins of Human Cognition*. Harvard University Press, Cambridge (MA).

Tomasello, M. (2005). *Constructing a Language: A Usage-Based Theory of Language Acquisition*. Harvard University Press, Cambridge (MA).

Tomczak, I. (2019). America's digital Messiah(s) of Detroit: Become human (2018). *New Horizons in English Studies*, 4: 158–172.

Tooby, J. and Cosmides, L., editors (2000). *Evolutionary Psychology: Foundational Papers*. Cambridge University Press, Cambridge (UK).

Tripp, B. P. (2017). Similarities and differences between stimulus tuning in the inferotemporal visual cortex and convolutional networks. In *International Joint Conference on Neural Networks*, pages 3551–3560.

Trischler, A., Ye, Z., Yuan, X., He, J., Bachman, P. and Suleman, K. (2016). A parallel-hierarchical model for machine comprehension on sparse data. *arXiv*, abs/1603.08884.

Trollope, A. (1876). *The Prime Minister*. Chapman and Hall, London.

Tsotsos, J. (1992). Motion understanding systems. In Sood, A. K. and Wechsler, H., editors, *Active Perception and Robot Vision*, pages 1–22. Springer-Verlag, Berlin.

Turing, A. (1936). On computable numbers, with an application to the Entscheidungsproblem. *Proceedings of the London Mathematical Society*, 42: 230–265.

Turing, A. (1948a). Intelligent machinery. Technical report, National Physical Laboratory, London. Raccolto in Ince, D. C. (ed.) *Collected Works of A. M. Turing: Mechanical Intelligence*, Edinburgh University Press, 1969.

Turing, A. (1948b). Intelligent machinery. Technical report, National Physical Laboratory, London. Reprinted in Ince, D. C. (ed.) *Collected Works of A. M. Turing: Mechanical Intelligence*, Elsevier Science Publishers, 1992.

Turing, A. (1950). Computing machinery and intelligence. *Mind*, 59: 433–460.

Van Essen, D. C. (2003). Organization of visual areas in macaque and human cerebral cortex. In Chalupa, L. and Werner, J., editors, *The Visual Neurosciences*. MIT Press, Cambridge (MA).

Van Essen, D. C. and DeYoe, E. A. (1994). Concurrent processing in the primate visual cortex. In Gazzaniga, M. S., editor, *The Cognitive Neurosciences*. MIT Press, Cambridge (MA).

van Heijenoort, J., editor (1967). *From Frege to Gödel: A Source Book in Mathematical Logic 1879–1931*. Harvard University Press, Cambridge (MA).

VanRullen, R. (2017). Perception science in the age of deep neural networks. *Frontiers in Psychology*, 8: 142.

Varela, F. J., Thompson, E. and Rosch, E. (1991). *The Embodied Mind – Cognitive Science and Human Experience*. MIT Press, Cambridge (MA).

Vartanian, A. (1973). Man-machine from the Greeks to the computer. In Wiener, P. G., editor, *Dictionary of the History of Ideas – Selected Pivotal Ideas*. Charles Scribner's Sons, New York.

Vaswani, A., Shazeer, N., Parmar, N., Uszkoreit, J., Jones, L., Gomez, A. N., Kaiser, Ł. and Polosukhin, I. (2017). Attention is all you need. In *Advances in Neural Information Processing Systems*, pages 6000–6010.

Venn, J. (1866). *The Logic of Chance: An Essay on the Foundations and Province of the Theory of Probability, with Especial Reference to Its Application to Moral and Social Science*. MacMillan and Co., London.

Veselý, K., Ghoshal, A., Burget, L. and Povey, D. (2013). Sequence-discriminative training of deep neural networks. In *Conference of the International Speech Communication Association*, pages 2345–2349.

Vignolo, A., Noceti, N., Rea, F., Sciutti, A., Odone, F. and Sandini, G. (2017). Detecting biological motion for human–robot interaction: A link between perception and action. *Frontiers in Robotics and AI*, 4: Article 14.

Vignolo, A., Rea, F., Noceti, N., Sciutti, A., Odone, F. and Sandini, G. (2016). Biological movement detector enhances the attentive skills of humanoid robot iCub. In *International Conference on Humanoid Robots (Humanoids)*, pages 338–344.

Vinge, V. (1993). The coming technological singularity: How to survive in the post-human era. In *Proc. Vision 21: Interdisciplinary Science and Engineering in the Era of Cyberspace*, pages 11–22, Lewis Research Center. NASA.

Vinyals, O., Kaiser, L., Koo, T., Petrov, S., Sutskever, I. and Hinton, G. (2015). Grammar as a foreign language. *arXiv*, abs/1412.7449.

Vinyals, O., Toshev, A., Bengio, S. and Erhan, D. (2016). Show and tell: Lessons learned from the 2015 MSCOCO image captioning challenge. *IEEE Transaction on Pattern Analysis and Machine Intelligence*, 39: 652–663.

von der Malsburg, C. (1973). Self-organization of orientation sensitive cells in the striate cortex. *Kybernetic*, 14: 85–100.

von der Malsburg, C. (1979). Development of ocularity domains and growth behaviour of axon terminals. *Biological Cybernetics*, 32: 49–62.

von der Malsburg, C. (1995). Network self-organization in the ontogenesis of the mammalian visual system. In Zornetzer, S. F., Davis, J., Lau, C. and McKenna, T., editors, *An Introduction to Neural and Electronic Networks*, pages 447–462. Academic Press, New York. Second Edition.

von der Malsburg, C. and Willshaw, D. J. (1976). A mechanism for producing continuous neural mappings: Ocularity dominance stripes and ordered retino-tectal projections. *Experimental Brain Research*, 1: 463–469.

von der Malsburg, C. and Willshaw, D. J. (1977). How to label nerve cells so that they can interconnect in an ordered fashion. *Proceedings of the Natural Academy of Science USA*, 74: 5176–5178.

Von Mises, R. (1931). *Wahrscheinlichkeitsrechnung und ihre Anwendungen in der Statistik und theoretischen Physik*. Franz Deuticke, Leipzig (DE).

Šmídl, V. and Quinn, A. (2005). *The Variational Bayes Method in Signal Processing*. Springer-Verlag, Berlin.

Wallace, R. S. (2008). The anatomy of A.L.I.C.E. In Epstein et al. (2008), pages 181–210.

Wallis, G. and Rolls, E. (1997). Invariant face and object recognition in the visual system. *Progress in Neurobiology*, 51: 167–194.

Walsh, B. and Gottfredson, F. (1947). The man of tomorrow. *Mickey Mouse comic strip*, September 22-December 27. reprint in F. Gottfredson, 2016, *Walt Disney's Mickey Mouse, Vol. 9: Rise of the Rhyming Man*, Seattle (WA): Fantagraphics Books.

Wamba, S. F., Bawack, R. E., Guthrie, C., Queiroz, M. M. and Carillo, K. D. A. (2021). Are we preparing for a good AI society? A bibliometric review and research agenda. *Technological Forecasting & Social Change*, 164: 120482.

Wang, A., Pruksachatkun, Y., Nangia, N., Singh, A., Michael, J., Hill, F., Levy, O. and Bowman, S. R. (2019a). Superglue: A stickier benchmark for general-purpose language understanding systems. In *Advances in Neural Information Processing Systems*, pages 3261–3275.

Wang, A., Singh, A., Michael, J., Hill, F., Levy, O. and Bowman, S. R. (2019b). GLUE: A multi-task benchmark and analysis platform for natural language understanding. In *International Conference on Learning Representations*.

Ward, M. (2018). *Seeming Human: Artificial Intelligence and Victorian Realist Character*. Ohio State University Press, Columbus (OH).

Warrington, E. K. (1975). The selective impairment of semantic memory. *Quarterly Journal of Experimental Psychology*, 27: 635–657.

Warrington, E. K. and Shallice, T. (1984). Category specific semantic impairments. *Brain*, 107: 829–853.

Warwick, K. and Shah, H. (2016). *Turing's Imitation Game – Conversations with the Unknown*. Cambridge University Press, Cambridge (UK).

Weber, M. (1921). *Wirtschaft und Gesellschaft. Grundrißder verstehenden Soziologie*. Mohr, Tübingen. English translation of Vol. 1 by Talcott Parsons, in *The Theory of Social and Economic Organization*, New York: Oxford University Press, 1947.

Weiss, R. (1969). *The American Myth of Success: From Horatio Alger to Norman Vincent Peale*. University of Illinois Press, Urbana (IL).

Weizenbaum, J. (1966). Eliza – A computer program for the study of natural language communication between man and machine. *Communications of the Association for Computing Machinery*, 9: 36–45.

Wells, A. J. (2004). Cognitive science and the Turing machine: an ecological perspective. In Teuscher (2004), pages 271–294.

Werbos, P. (1974). *Beyond Regression: New Tools for Prediction and Analysis in the Behavioral Sciences*. PhD thesis, Harvard University.

Werbos, P. (1994). *The Roots of Backpropagation: From Ordered Derivatives to Neural Networks*. John Wiley, New York.

Westermann, G. (1998). Emergent modularity and U-shaped learning in a constructivist neural network learning the English past tense. In Gernsbacher, M. and Derry, S., editors, *Proceedings of the 20th Annual Conference of the Cognitive Science Society*, pages 1130–1135.

Westermann, G. and Plunkett, K. (2007). Connectionist models of inflection processing. *Lingue e Linguaggio*, 6: 291–311.

Whitehead, A. N. and Russell, B. (1910, 1912, 1913). *Principia Mathematica*. Chicago University Press, Chicago (IL). 3 Vols., Second Edition, 1925 (Vol. 1), 1927 (Vols. 2, 3).

Wickelgren, W. (1969). Context sensitive coding, associative memory, and serial order in (speech) behavior. *Psychological Review*, 76: 1–15.

Wiener, N. (1948). *Cybernetics, or Control and Communication in the Animal and the Machine*. MIT Press, Cambridge (MA).

Wild, J. (1955). *The Challenge of Existentialism*. Indiana University Press, Bloomington (IN).

Wilkes, D. R. and Tsotsos, J. (1993). Behaviours for active object recognition. In *Intelligent Robots and Computer Vision*, pages 225–239. SPIE.

Williams, G. (1992). *Natural Selection: Domains, Levels, and Challenges*. Oxford University Press, Oxford (UK).

Willshaw, D. J. and von der Malsburg, C. (1976). How patterned neural connections can be set up by self-organization. *Proceedings of the Royal Society of London*, B194: 431–445.

Wilson, K. G. and Kogut, J. (1974). The renormalization group and the ε expansion. *Physics Reports*, 12: 75–199.

Wimsatt, W. K. (1954). *The Verbal Icon: Studies in the Meaning of Poetry*. University Press of Kentucky, Louisville (KY).

Winograd, T. (1972). *Understanding Natural Language*. Academic Press, New York.

Winograd, T. and Flores, F. (1987). *Understanding Computers and Cognition – A New Foundation for Design*. Addison Wesley, Reading (MA).

Woessner, M. (2011). *Heidegger in America*. Cambridge University Press, Cambridge (UK).

Wright, J. and Mourshed, M. (2009). Geometric optimization of fenestration. In *Proceedings of Building Simulation*, pages 920–927.

Wu, E. and Liu, Y. (2008). Emerging technology about GPGPU. In *IEEE Asia Pacific Conference on Circuits and Systems*, pages 618–622.

Yamins, D. L. K., Honga, H., Cadieua, C. F., Solomon, E. A., Seibert, D. and DiCarlo, J. J. (2014). Performance-optimized hierarchical models predict neural responses in higher visual cortex. *Proceedings of the Natural Academy of Science USA*, 23: 8619–8624.

Yang, X.-S. and He, X. (2015). Swarm intelligence and evolutionary computation: Overview and analysis. In Yang, X. -S., editor, *Recent Advances in Swarm Intelligence and Evolutionary Computation*, pages 1–23. Springer, Berlin.

Yang, Y., Tarr, M. J. and Elissa M Aminoff, R. E. K. (2019). Exploring spatio–temporal neural dynamics of the human visual cortex. *Human Brain Mapping*, 40: 4213–4238.

Yang, Z., Wu, W., Xu, C., Liang, X., Bai, J., Wang, L., Wang, W. and Li, Z. (2020). StyleDGPT: Stylized response generation with pre-trained language models. In *Findings of the Association for Computational Linguistics: EMNLP 2020*, pages 1548–1559.

Zeng, Y., Zhao, Y., Bai, J. and Xu, B. (2018). Toward robot self-consciousness (II): Brain-inspired robot bodily self model for self-recognition. *Cognitive Computation*, 10: 307–320.

Zeppi, A. and Blokpoel, M. (2017). Mindshaping the world can make mindreading tractable: Bridging the gap between philosophy and computational complexity analysis. In *Annual Meeting of the Cognitive Science Society*, pages 1418–1423.

Zhabotinsky, A. (1964). Periodical process of oxidation of malonic acid solution. *Biophysics*, 56: 178–194.

Zhang, S., Dinan, E., Urbanek, J., Szlam, A., Kiela, D. and Weston, J. (2018). Personalizing dialogue agents: I have a dog, do you have pets too? *arXiv*, abs/1801.07243.

Zhang, Y., Sun, S., Galley, M., Chen, Y.-C., Brockett, C., Gao, X. et al. (2020). DialoGPT: Large-scale generative pre-training for conversational response generation. *arXiv*, abs/1911.00536.

Zhou, J., Cao, Y., Wang, X., Li, P. and Xu, W. (2016). Deep recurrent models with fast-forward connections for neural machine translation. *Transactions of the Association for Computational Linguistics*, 4: 371–383.

Zhu, H., Wei, F., Qin, B. and Liu, T. (2018). Hierarchical attention flow for multiple-choice reading comprehension. In *AAAI Conference on Artificial Intelligence*, pages 6077–6084.

Ziman, J. (2000). *Real Science – What It Is and What It Means*. Cambridge University Press, Cambridge (UK).

Zimbardo, P. G. (2007). *The Lucifer Effect: How Good People Turn Evil*. Random House, New York.

Index

Printed in the United States
by Baker & Taylor Publisher Services